Lecture Notes in Artificial Intelligence   8105

Subseries of Lecture Notes in Computer Science

LNAI Series Editors

Randy Goebel
   University of Alberta, Edmonton, Canada
Yuzuru Tanaka
   Hokkaido University, Sapporo, Japan
Wolfgang Wahlster
   DFKI and Saarland University, Saarbrücken, Germany

LNAI Founding Series Editor

Joerg Siekmann
   DFKI and Saarland University, Saarbrücken, Germany

T0202623

Iryna Gurevych   Chris Biemann
Torsten Zesch (Eds.)

Language Processing and Knowledge in the Web

25th International Conference, GSCL 2013
Darmstadt, Germany, September 25-27, 2013
Proceedings

 Springer

Volume Editors

Iryna Gurevych
Ubiquitous Knowledge Processing Lab
Department of Computer Science, Technische Universität Darmstadt, Darmstadt,
Germany and Ubiquitous Knowledge Processing Lab, German Institute
for International Educational Research (DIPF), Frankfurt am Main, Germany
E-mail: gurevych@ukp.informatik.tu-darmstadt.de

Chris Biemann
FG Language Technology
Department of Computer Science, Technische Universität Darmstadt
Darmstadt, Germany
E-mail: biem@cs.tu-darmstadt.de

Torsten Zesch
Ubiquitous Knowledge Processing Lab
Department of Computer Science, Technische Universität Darmstadt, Darmstadt,
Germany and Ubiquitous Knowledge Processing Lab, German Institute
for International Educational Research (DIPF) Frankfurt am Main, Germany
E-mail: zesch@ukp.informatik.tu-darmstadt.de

ISSN 0302-9743   e-ISSN 1611-3349
ISBN 978-3-642-40721-5   e-ISBN 978-3-642-40722-2
DOI 10.1007/978-3-642-40722-2
Springer Heidelberg New York Dordrecht London

Library of Congress Control Number: 2013946726

CR Subject Classification (1998): I.2, H.3, H.4, F.1, I.5, I.4, C.2

LNCS Sublibrary: SL 7 – Artificial Intelligence

© Springer-Verlag Berlin Heidelberg 2013

Typesetting: Camera-ready by author, data conversion by Scientific Publishing Services, Chennai, India

Printed on acid-free paper

Springer is part of Springer Science+Business Media (www.springer.com)

Dedication

This volume is dedicated to the memory of Prof. Wolfgang Hoeppner, who passed away on June 4, 2012.

For many years, Wolfgang Hoeppner was a member and coordinator of the Scientific Board of the German Society for Computational Linguistics and Language Technology (GSCL). He contributed greatly to GSCL and the interdisciplinary development of GSCL at the intersection of knowledge discovery, human–computer interaction, and language technology.

Iryna Gurevych, Chris Biemann, and Torsten Zesch
Co-editors

Preface

The International Conference of the German Society for Computational Linguistics and Language Technology (GSCL 2013) was held in Darmstadt, Germany, during September 25–27, 2013. The meeting brought together an international audience from Germany and other countries. The conference's main theme was "Language Processing and Knowledge in the Web."

Language processing and knowledge in the Web has been an area of great and steadily increasing interest within the language processing and related communities over the past years. Both in terms of academic research and commercial applications, the Web has stimulated and influenced research directions, yielding significant results with impact beyond the Web itself. Thus, the conference turned out to be a very useful forum in which to highlight the most recent advances within this domain and to consolidate the individual research outcomes.

The papers accepted for publication in the present Springer volume address language processing and knowledge in the Web on several important dimensions, such as computational linguistics, language technology, and processing of unstructured textual content in the Web.

About one third of the papers are dedicated to fundamental computational linguistics research in multilingual settings. On the one hand, the work deals with different languages, such as German, Manipuri, or Chinese. On the other hand, it deals with a wide range of computational linguistics tasks, such as word segmentation, modeling compounds, coreference resolution, word sense annotation, named entity recognition, or lexical-semantic processing.

The second third of papers address a wide range of language technology tasks, such as construction of a new error tagset for an Arabic learning corpus and prediction of cause of death from verbal autopsy text. Two papers deal with different aspects of machine translation. An evaluation of several approaches to sentiment analysis is the subject of another contribution. Last but not least, one article deals with dependency-based algorithms in question answering for Russian.

The third portion of the papers presented in this volume deals with processing of unstructured textual content in the Web. An important issue is the construction of Web corpora for computational research. One paper presents a tool for creating tailored Twitter corpora, while another describes the construction of a corpus of parsable sentences from the Web. Optimizing language processing components to work on noisy Web content is the subject of several papers. Finally, one contribution exploits Wikipedia as a knowledge resource for topic modeling, and another presents a novel summarization algorithm for community-based question-answering services.

In summary, the GSCL 2013 conference clearly demonstrated the recent advances in language processing research for processing the textual content in the

Web. It also showed that Web corpora can be effectively employed as a resource in language processing. A particular property of the Web is its multilinguality, which is reflected in a significant number of papers dealing with languages other than English and German published in the present volume.

We would like to sincerely thank the Organizing Committee of GSCL 2013 and the reviewers for their hard work, the invited speakers for their inspiring contributions to the program, the sponsors and funding agencies for their financial contributions, and Tristan Miller for his technical assistance in compiling the final volume. We also express our gratitude to the Hessian LOEWE research excellence program and to the Volkswagen Foundation for funding the conference organizers as part of the research center "Digital Humanities" (Chris Biemann) and the Lichtenberg Professorship Program under grant № I/8280 (Iryna Gurevych).

Iryna Gurevych
Chris Biemann
Torsten Zesch

Organization

GSCL 2013 was organized by the Ubiquitous Knowledge Processing (UKP) Lab of the Technische Universität Darmstadt's Department of Computer Science.

Organizing Committee

General Chair

Iryna Gurevych — Technische Universität Darmstadt and German Institute for International Educational Research (DIPF), Germany

Program Chairs

Chris Biemann — Technische Universität Darmstadt, Germany
Torsten Zesch — Technische Universität Darmstadt and German Institute for International Educational Research (DIPF), Germany

Workshops/Tutorials Chair

György Szarvas — Nuance Communications, Germany

Local Chairs

Christian M. Meyer — Technische Universität Darmstadt, Germany
Wolfgang Stille — Technische Universität Darmstadt, Germany

Program Committee

Abend, Omri — The Hebrew University of Jerusalem, Israel
Auer, Sören — Universität Leipzig, Germany
Bernhard, Delphine — Université de Strasbourg, France
Buitelaar, Paul — Digital Enterprise Research Institute, National University of Ireland, Ireland
Choudhury, Monojit — Microsoft Research, India
Cimiano, Philipp — Bielefeld University, Germany
Cysouw, Michael — Philipps-Universität Marburg, Germany
Dagan, Ido — Bar-Ilan University, Israel
De Luca, Ernesto William — University of Applied Sciences, Potsdam, Germany
de Melo, Gerard — International Computer Science Institute, USA
Dipper, Stefanie — Ruhr-Universität Bochum, Germany
Fellbaum, Christiane — Princeton University, USA

Frank, Annette   Heidelberg University, Germany
Girju, Roxana   University of Illinois at Urbana-Champaign,
                                  USA
Heid, Ulrich   University of Hildesheim, Germany
Heyer, Gerhard   Universität Leipzig, Germany
Hirst, Graeme   University of Toronto, Canada
Hoeppner, Wolfgang (†)   Universität Duisburg-Essen, Germany
Kozareva, Zornitsa   Information Sciences Institute, University of
                                  Southern California, USA
Lobin, Henning   Justus Liebig University Giessen, Germany
Lüdeling, Anke   Humboldt-Universität zu Berlin, Germany
Magnini, Bernardo   ITC-irst, Italy
Mahlow, Cerstin   University of Zurich, Switzerland
Manandhar, Suresh   University of York, UK
McCarthy, Diana   University of Sussex, UK
Mehler, Alexander   Goethe University Frankfurt am Main,
                                  Germany
Mihalcea, Rada   University of North Texas, USA
Miller, Tristan   Technische Universität Darmstadt, Germany
Mohammad, Saif   National Research Council Canada, Canada
Navigli, Roberto   Sapienza University of Rome, Italy
Nenkova, Ani   University of Pennsylvania, USA
Neumann, Günter   German Research Center for Artificial
                                  Intelligence (DFKI), Germany
Ng, Vincent   University of Texas at Dallas, USA
Padó, Sebastian   Heidelberg University, Germany
Palmer, Alexis   Saarland University, Germany
Poesio, Massimo   University of Essex, UK
Quasthoff, Uwe   Universität Leipzig, Germany
Rehm, Georg   German Research Center for Artificial
                                  Intelligence (DFKI), Germany
Riezler, Stefan   Heidelberg University, Germany
Schlangen, David   Bielefeld University, Germany
Schmidt, Thomas   Institut für Deutsche Sprache, Germany
Schmitz, Ulrich   University of Duisburg-Essen, Germany
Schröder, Bernhard   University of Duisburg-Essen, Germany
Stein, Benno   Bauhaus-Universität Weimar, Germany
Storrer, Angelika   Technische Universität Dortmund, Germany
Søgaard, Anders   University of Copenhagen, Denmark
Teich, Elke   Saarland University, Germany
Temnikova, Irina   University of Wolverhampton, UK
Wandmacher, Tonio   SYSTRAN, France
Witt, Andreas   Institut für Deutsche Sprache, Germany
Witte, René   Concordia University, Canada
Wolff, Christian   Universität Regensburg, Germany

Big Data and Text Analytics

Hans Uszkoreit

Saarland University, Germany
German Research Center for Artificial Intelligence (DFKI),
Germany

Abstract. Text analytics is faced with rapidly increasing volumes of language data. In our talk we will show that big language data are not only a challenge for language technology but also an opportunity for obtaining application-specific language models that can cope with the long tail of linguistic creativity. Such models range from statistical models to large rule systems. Using examples from relation/event extraction we will illustrate the exploitation of large-scale learning data for the acquisition of application specific syntactic and semantic knowledge and discuss the achieved improvements of recall and precision.

Biography: Hans Uszkoreit is Professor of Computational Linguistics and—by cooptation—of Computer Science at Saarland University. At the same time he serves as Scientific Director at the German Research Center for Artificial Intelligence (DFKI) where he heads the DFKI Language Technology Lab. He has more than 30 years of experience in language technology which are documented in more than 180 international publications. Uszkoreit is Coordinator of the European Network of Excellence META-NET with 60 research centers in 34 countries and he leads several national and international research projects. His current research interests are information extraction, atomatic translation and other advanced applications of language and knowledge technologies as well as computer models of human language understanding and production.

Distributed Wikipedia LDA

Massimiliano Ciaramita

Google Research
Zurich, Switzerland

Abstract. When someone mentions Mercury, are they talking about the planet, the god, the car, the element, Freddie, or one of some 89 other possibilities? This problem is called disambiguation, and while it's necessary for communication, and humans are amazingly good at it, computers need help. Automatic disambiguation is a long standing problem and is the focus of much recent work in natural language processing, web search and data mining. The surge in interest is due primarily to the availability of large scale knowledge bases such as Wikipedia and Freebase which offer enough coverage and structured information to support algorithmic solutions and web-scale applications. In this talk I will present recent work on the disambiguation problem based on a novel distributed inference and representation framework that builds on Wikipedia, Latent Dirichlet Allocation and pipelines of MapReduce.

Biography: Massimiliano Ciaramita is a research scientist at Google Zurich. Previously he has worked as a researcher at Yahoo! Research and the Italian National Research Council. He did his undergraduate studies at the University of Rome "La Sapienza" and obtained ScM and PhD degrees from Brown University. His main research interests involve language understanding and its applications to search technologies. He has worked on a wide range of topics in natural language processing and information retrieval, including disambiguation, acquisition, information extraction, syntactic and semantic parsing, query analysis, computational advertising and question answering. He co-teaches (with Enrique Alfonseca) "Introduction to Natural Language Processing" at ETH Zurich.

Multimodal Sentiment Analysis

Rada Mihalcea

Department of Computer Science and Engineering
University of North Texas, USA

Abstract. During real-life interactions, people are naturally gesturing and modulating their voice to emphasize specific points or to express their emotions. With the recent growth of social websites such as YouTube, Facebook, and Amazon, video reviews are emerging as a new source of multimodal and natural opinions that has been left almost untapped by automatic opinion analysis techniques. One crucial challenge for the coming decade is to be able to harvest relevant information from this constant flow of multimodal data. In this talk, I will introduce the task of multimodal sentiment analysis, and present a method that integrates linguistic, audio, and visual features for the purpose of identifying sentiment in online videos. I will first describe a novel dataset consisting of videos collected from the social media website YouTube and annotated for sentiment polarity at both video and utterance level. I will then show, through comparative experiments, that the joint use of visual, audio, and textual features greatly improves over the use of only one modality at a time. Finally, by running evaluations on datasets in English and Spanish, I will show that the method is portable and works equally well when applied to different languages.

Biography: Rada Mihalcea is an Associate Professor in the Department of Computer Science and Engineering at the University of North Texas. Her research interests are in computational linguistics, with a focus on lexical semantics, graph-based algorithms for natural language processing, and multilingual natural language processing. She serves or has served on the editorial boards of the journals of *Computational Linguistics, Language Resources and Evaluation, Natural Language Engineering, Research in Language in Computation, IEEE Transations on Affective Computing*, and *Transactions of the Association for Computational Linguistics*. She was a program co-chair for the Conference of the Association for Computational Linguistics (2011), and the Conference on Empirical Methods in Natural Language Processing (2009). She is the recipient of a National Science Foundation CAREER award (2008) and a Presidential Early Career Award for Scientists and Engineers (2009).

Table of Contents

Reconstructing Complete Lemmas
for Incomplete German Compounds

Noëmi Aepli and Martin Volk

University of Zurich
noemi.aepli@uzh.ch, volk@cl.uzh.ch

Abstract. This paper discusses elliptical compounds, which are frequently used in German in order to avoid repetitions. This phenomenon involves truncated words, mostly truncated compounds. These words pose a challenge in PoS tagging and lemmatization, which often leads to unknown or incomplete lemmas. We present an approach to reconstruct complete lemmas of truncated compounds in order to improve subsequent language technology or corpus linguistic applications. Results show an f-measure of 95.6% for the detection of elliptical compound patterns and 86.4% for the correction of compound lemmas.

Keywords: Elliptical compounds, decompounding, corpus annotation.

1 Introduction

Many languages use elliptical constructions in order to avoid repetition and to allow concise wording. In this paper we focus on elliptical compounds in German coordination constructions. We use the term "elliptical compound"[1] to refer to truncated words like *Schnee-* in coordinated constructions such as *Schnee- und Lawinenforschung* (snow [research] and avalanche research). These truncated words typically end with a hyphen, which stands for the last part of a full compound following the conjunction. The full coordination in the example would be *Schneeforschung und Lawinenforschung*.

Since we want to have access to all compounds in our corpus in a unified way, we are interested in resolving the hyphen references of truncated words. Current Part-of-Speech (PoS) taggers for German usually assign the dummy tag TRUNC to such elliptical compounds (Thielen et al., 1999). Most current lemmatizers work without regard to context and therefore they cannot assign the full compound as lemma for a truncated word. They either use the word form as lemma, or they opt for the dummy lemma "unknown". Since incomplete lemmas are a stumbling block for any type of machine processing, our goal is to overcome this restriction and to deliver a full lemma for a truncated word based on an analysis of the coordination construction.

---

[1] We are aware of the fact that not all the cases matching our patterns are compounds, e.g. *jenseits* in *dies- und jenseits* (on this [side] and on the other side) is not a compound. However, compounds are by far the most frequent and typical.

I. Gurevych, C. Biemann, and T. Zesch (Eds.): GSCL 2013, LNAI 8105, pp. 1–13, 2013.
© Springer-Verlag Berlin Heidelberg 2013

Towards this goal we have collected the most typical coordination patterns in German that involve elliptical compounds. We have developed a program that, when triggered by a truncated word, determines the coordination pattern, splits the full compound of the construction into its elements, and uses the last segment of this full compound to generate the full lemma of the truncated word. The freedom in creating elliptical compounds in German and the resulting diversity of constructions turn this into an interesting challenge.

We have developed and tested our lemma reconstructor on the German part of the Text+Berg corpus, a large collection of texts from the Swiss Alpine Club (Volk et al., 2010). However, the coordination patterns and the methodology are applicable to other corpora as well.

In the next section, we briefly introduce some related work, and we describe the linguistic phenomenon of elliptical compounds. In section 3 we introduce the Text+Berg corpus and its characteristics. In section 4, we give an overview of the general system architecture, illustrated with examples. Section 6 presents our evaluation of the lemma reconstructor and mentions some problematic cases.

2   Elliptical Compounds

The linguistic properties of compounds are widely studied and some of these works include sidesteps on elliptical compounds with hyphens. Several types of hyphens exist; one is used to break single words into parts if the word continues in the following line, another to join separate words into one word. For our work, the third usage is important; the suspended hyphen which marks the truncated word (Bredel, 2008). Srinivasan (1993) sees the suspended hyphen as a morpheme placeholder whereas he describes the other two usages as "word breaks" since they break connected chains of letters.

According to the official spelling rules 98[2] and the Duden dictionary rule 31[3] a suspended hyphen can replace the lexeme which the compounds have in common. This can be the first and/or the last lexeme:

- first lexeme: *Bergkameradschaft und -hilfe* (moutain fellowship and [mountain] help)
- last lexeme: *laut- und spurlos* (sound[less] and traceless)
- first and last lexeme: *Sonnenauf- und -untergang* (sun[rise] and [sun]set)

The truncated word is described by Eisenberg (2004) as a separate syntactic base form; the "word rest". This indicates that something has been omitted, which, however, can be regained in the surroundings.

The first challenge is to find all the patterns of such compounds, the second is the correct word segmentation. The latter task is complicated by the freedom of merging words as a common type of word formation in German. Combining word stems into one compound leads to a huge increase of vocabulary and

---

[2] canoo.net/services/GermanSpelling/Amtlich/Inter-punktion/pgf98.html, 2.10.2012.

[3] www.duden.de/sprachwissen/rechtschreibregeln/binde-strich#K31, 2.10.2012.

hence to a sparse data problem (Holz and Biemann, 2008). This is also shown by some word formations occurring in the Text+Berg corpus, e.g. *Edelweiss-romantik* (Edelweiss romance), *Akrobatentänzerverein* (acrobat dancer club) or *Wegwerflandschaften* (disposable landscapes).

In order to obtain an overview of the frequency of the elliptical compound phenomenon we investigated the TIGER treebank (Brants et al., 2002), a collection of 50,474 German syntax trees corresponding to 888,299 tokens. In this corpus we found 1226 sentences (about 2% of all sentences) with a total of 1362 tokens which are tagged as TRUNCated words. Figure 1 shows an example with two such truncated words *Natur-, Umwelt-*. Note the reduced lemmas for these words. Since the full compound *Lebensschutz* is not given with segmentation boundaries, it is far from trivial to reconstruct the complete lemmas for the truncated words.

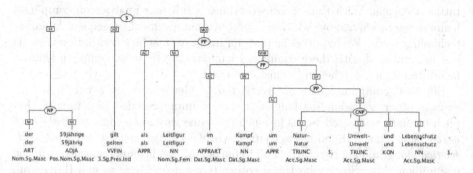

Fig. 1. Example sentence from the TIGER treebank with two truncated words

The statistics show that the TIGER treebank amongst others includes

- 6 TRUNC tokens with digits (e.g. *16- und 17jährigen* (16[-year-old] and 17-year-old))
- 110 TRUNC tokens that start with a lower case letter (e.g. *in- und ausländischen* (domestic and foreign), *bi- als auch multilaterale* (bi[lateral] and multilateral), *mittel- bis langfristig* (medium-[term] to long-term))
- 28 TRUNC tokens without hyphen whose classification is dubious (e.g. *des (ab/TRUNC) laufenden Jahrhunderts* (of the (expiring) continuous decade), *dem nachkolonialen/TRUNC und Nachkriegs-Vietnam* (the post colonial and post war Vietnam), *der Beteiligungs/TRUNC AG* (the holding corporation), *Fast/TRUNC Food*)

Words with initial hyphens are not tagged as TRUNC in the TIGER corpus. They are tagged as if the hyphens were not present, mostly they are regular nouns (NN). We found 47 words which start with a hyphen and are part of a coordinated nominal phrase, which indicates that they are most probably part of such a construction. Some examples:

- *Desintegrations-Ängste und -Erfahrungen/NN* (disintegration fears and [disintegration] experiences)

– *Bodenverwaltungs- und -verwertungsgesellschaft/NN* (land administration [corporation] and [land] usage corporation)
– *weder Arbeitslosengeld noch -hilfe/NN* (neither unemployment benefit nor [unemployment] aid)

In the TIGER corpus the lemmas of the truncated words correspond to the first lexeme of the compound. This means the hyphen is omitted, thus the truncated compound of *vier- bis fünfhundert* (four [hundred] to five hundred) is annotated with the lemma *vier* (four). If there is a linking element (*-s-* or *-es-*), it is also omitted which means that the truncated word of *Bundes- und Reichsbahn* (federal [railway] and state railway) is annotated with the lemma *Bund* (confederation).

The aforementioned examples show how much information gets lost, which implies that our completion of elliptical compounds is a useful step when annotating a corpus. With these reconstructions, a full text analysis on compound lemmas can be carried out which will improve the results of subsequent language technology tasks. We hypothesize that, for instance, machine translation systems can be improved with these completed lemmas. In any case, complete lemmas facilitate corpus searches in linguistic research.

Elliptical compounds are not restricted to German. They occur in similar ways in other compounding languages. For example, Swedish also uses the hyphen to mark truncated words in much the same way as German (e.g. *olje- och gasverksamheten* (oil and gas activities), *Nord- och Sydamerika, lång- och kortfristiga* (long and short-term)). The frequency of last lexeme truncations in the one million token Stockholm-Umeå corpus (Gustafson-Capková and Hartmann, 2006) is on the same order of magnitude as in the German TIGER corpus. In contrast, truncated words with hyphens for first lexeme omission are very rare in Swedish (e.g. *tillverkningsprocesser och -metoder* (production processes and methods)). Our approach to reconstruct the lemmas of incomplete closed compounds also has a correspondence for open compounds. Unlike languages like German, where compounds are usually written as one word, in English solid compounds are very rare. However, open compounds, where the constituents of compounds are separated by white spaces, are commonly used (Rakić, 2009). Consider the coordination construction *oil and gas activities*. In order to interpret this coordination correctly, a system has to detect that both *oil* and *gas* are modifiers of *activities*. The resolution thus amounts to disambiguation of modifiers in complex coordinations.

3   The Text+Berg Corpus

We have digitized all yearbooks of the Swiss Alpine Club from 1864 until today. Most of the articles are in French and German, a few in English, Italian and Romansh. We scanned the texts and used OCR (optical character recognition) software to convert the scanned images into text (Volk et al., 2011). We then structured the text by identifying article boundaries based on manually corrected tables of contents. We tokenized and split each text into sentences and determined the language of each sentence automatically. This procedure allows, amongst others, the recognition of German quotations in French articles and vice versa.

Tokenization of the corpus was a major undertaking because of the spelling idiosyncrasies over the 150-year time span and the different languages. In particular, apostrophes and hyphens caused problems. Hyphens that result from end-of-line word breaks were eliminated by checking whether the form without hyphen was clearly more frequent in the corpus than the form with the hyphen. Hyphenated compounds were left as one token (e.g. *SAC-Hütte*), but French combinations of verb + pronoun were split (e.g. *viennent-ils → viennent + -ils*) in order to facilitate PoS tagging. Word-final hyphens were left intact as they allow us to identify truncated words.

After tokenization the corpus was PoS-tagged and lemmatized with different parameter files for the TreeTagger (Schmid, 1994) for English, French, German and Italian. We added missing lemmas for German and French based on other dictionaries. Sentences in Romansh were not annotated since there is no PoS tagger and lemmatizer for this language. Subsequently special modules were run for named entity recognition (geographical names and person names) and for alignment between the French-German translated parts of the corpus. All the information is stored in XML files.

The Text+Berg corpus currently consists of 22.5 million tokens in German and 21.5 million tokens in French, out of which about 5 million tokens are translations (i.e. parallel texts). The Italian part has 0.3 million tokens, English and Romansh have less than 100,000 tokens each. The corpus is freely available for research purposes.

4   Architecture of Our Lemma Reconstructor

As input, our Python script takes a corpus file in XML and two lists containing word frequencies from our corpus and word segmentations from a morphological analyzer. The input XML file represents a typical elliptical compound construction as depicted in figure 2. Each token is annotated with a unique identifier, a PoS tag, and a lemma. After the reconstruction both the elliptical compound and the full compound have complete lemmas which include the segmentation boundary markers (see figure 2).

The task of completing lemmas can roughly be divided into two steps. First, our lemma reconstructor looks for patterns that contain elliptical compounds, and second, found patterns are modified in order to complete the lemma of the truncated word with the missing part.

4.1   Pattern Matching

The main function runs through the parsed XML file looking for patterns. If a pattern matches, the task is forwarded to several modules in order to analyze the compound construction and identify the missing part.

```
--- Before Lemma Reconstruction ---
<w n="8-398-7" pos="TRUNC" lemma="Kuh-">Kuh-</w>
<w n="8-398-8" pos="KON"   lemma="und">und</w>
<w n="8-398-9" pos="NN"    lemma="Ziegenherde">Ziegenherden</w>

--- After Lemma Reconstruction ---
<w n="8-398-7" pos="TRUNC" lemma="Kuh#herde">Kuh-</w>
<w n="8-398-8" pos="KON"   lemma="und">und</w>
<w n="8-398-9" pos="NN"    lemma="Ziegen#herde">Ziegenherden</w>
```

Fig. 2. XML example before and after our lemma reconstruction program

The German part of the Text+Berg corpus is part-of-speech annotated with the Stuttgart Tübingen Tagset (Thielen et al., 1999), which includes a PoS tag for truncated words: *TRUNC*. We use the part-of-speech information in order to define the patterns. We first investigated the TIGER Corpus and set the most frequent patterns as a starting point. In a next step, we applied them to the Text+Berg corpus and redefined, corrected, as well as extended them continuously. For some patterns more than one solution is possible. In such cases, all the possible solutions are generated and with the help of word frequencies the most likely solution is chosen. There are eleven specified patterns[4], some of which have several solution variants.

(1) TRUNC1 + \$, + TRUNC2 + \$, + TRUNC3 + KON + NN/NE/ADJA
   ⇒ TRUNC1 + WORD_L & TRUNC2 + WORD_L & TRUNC3 + WORD_L
   → *Kondordia-, Gleckstein-, Dossen- und Gaulihütte*
   (Konkordia [Hut], Gleckstein [Hut], Dossen [Hut] and Gauli **Hut**)

(2) TRUNC1 + \$, + TRUNC2 + KON + NN/NE/ADJA
   ⇒ TRUNC1 + WORD_L & TRUNC2 + WORD_L
   → *Wind-, Niederschlags- und Temperatur-messungen*
   (wind [measurements], rain [measurements] and temperature **measurements**)

(3) TRUNC + KON + -WORD
   ⇒ TRUNC + -WORD_L & TRUNC_L + -WORD
   → *Schieferschutt- und -plattenhang* (**slate** scree [slope] and [slate] slab **slope**)

(4) TRUNC + KON + APPR + ART + NN/NE
   (a) TRUNC + NN_L
       → *Nord- und auf der Ostseite* (north [side] and on the east **side**)
   (b) TRUNC + APPR_L
       → *dies- und jenseits* (this [side] and on the other **side**)

(5) TRUNC + KON + ADJ/ADV/VVFIN/VVIZU/ VVPP/CARD/APPR + !NN/ NE
   ⇒ TRUNC + WORD_L
   → *hinauf- und hinuntergeklettert* ([climbed] up and **climbed** down)

---

[4] / stands for *or*, ! for *not*, F/L for the first/last lexeme of a word, (a), (b), (c) for different solution variants, ⇒ implies that there is only one specified solution.

(6) TRUNC + KON + ADJA/CARD/ADV + NN/NE
   (a) TRUNC + NN
     → *Alpin- und sonstige **Rucksäcke*** (alpine [backpacks] and other **backpacks**)
   (b) TRUNC + ADJ_L
     → *hilf- und gnaden**reichen** Mutter* (help[ful] and merci**ful** mother)
   (c) TRUNC + NN_L
     → *Sessel- und vier Ski**lifts*** (chair [lifts] and four ski **lifts**)

(7) TRUNC + KON + ART/APPR/APPRART + NN/NE
   ⇒ TRUNC + NN_L
   → *Fels- oder durch Schutt**massen*** (rock [material] or through scree **material**)

(8) TRUNC + KON + ART/CARD/ADV/APPR/APPR-ART + ADJA + NN/NE
   (a) TRUNC + NN
     → *Languard- und die angrenzenden **Gebiete***
     (Languard [areas] and the bordering **areas**)
   (b) TRUNC + NN_L
     → *Walen- und dem Oberen Zürich**see***
     ([Lake] Walen and the Upper *Lake* Zurich)
   (c) TRUNC + ADJA_L
     → *französisch- und der deutsch**sprachigen** Schweiz*
     (French [speaking] and German **speaking** Switzerland)

(9) TRUNC + KON + NN
   ⇒ TRUNC + NN_L
   → *Eis- und Schnee**wänden*** (ice [walls] and snow **walls**)

(10) TRUNC + APPRART + NN/NE/ADJA
   ⇒ TRUNC + NN_L
   → *Nord- zum Süd**gipfel*** (north [peak] to the south **peak**)

(11) WORD1 + KON + -WORD2
   ⇒ WORD1_F + -WORD2
   → ***Berg**steiger und -führer* (**mountain** climber and [mountain] guide)

4.2 Compound Analysis, Solution Generation and Decision

If, for example, *Schnee- und Lawinenforschung* (snow [research] and avalanche research) has been found matching the pattern *(9) TRUNC + KON + NN*, the construction is split for further processing. First, the complete compound *Lawinenforschung* is analysed by Gertwol[5], a wide-coverage morphology system for German. Gertwol de-compounds all words where all segments are known to the system. In case of multiple segmentation variants we use the algorithm proposed by Volk (1999) for disambiguation. Gertwol provides an analysis with four different segmentation symbols[6] to differentiate between different

---

[5] www.lingsoft.fi
[6] www2.lingsoft.fi/doc/gertwol/intro/segment.html, 2.10.2012.

word-internal boundaries.[7] In our example, Gertwol delivers the segmentation *Lawine\n#forsch~ung*. We disregard the linking element and the suffix marker, and so this compound consists of only two parts, i.e. *Lawine* and *forschung*. By recognizing the strong compound boundary (#), we pick the second part to fill the lemma of the truncated word *Schnee-*.

If the input word is unknown to Gertwol, we try to segment it with the help of words from our corpus. In addition we use their frequencies in our corpus in order to determine the most likely split. This is done by splitting the compound in every possible way into two parts, so that there are at least three characters left on the left and the right side. The truncated word is then concatenated with each possible right part and the most frequent word, according to our corpus, is taken as the solution. In the example *Kuh- und Yakherden* ([herds of] cows and herds of yak) we split the word *Yakherden* and generate the variants *Kuh#herden*, *Kuh#erden*, *Kuh#rden* and *Kuh#den* with word parts. Since *Kuhherden* occurs 11 times in our corpus and all the alternatives do not occur at all, we select this compound for reconstruction and adopt its lemma *Kuh#herde*.

If, in another case, Gertwol provides the information that the input word has more than one strong boundary (#), like for example *Schnee#schuh#touren* in *Ski- und Schneeschuhtouren* (ski [touring] and snowshoe touring), we generate all possible alternatives for the reconstructed lemma by concatenating the truncated word with each of the possible missing parts; *Ski#schuhtour* and *Ski#tour*. Again the corpus frequencies 0 vs. 440 enable us to select the correct lemma *Ski#tour*. We recently realized that this procedure could also help to determine the internal structure of complex compounds, since it will predict that *(Schnee#schuh)#touren* is a more likely interpretation than *Schnee#(schuh#-touren)*. We have not yet explored this idea any further.

5   Manual Error Analysis

During development and testing we encountered typical difficulties with elliptical compounds. In this section we present some of the errors along with ideas on how to solve these problems.

Decision Problem: A main drawback of our approach are cases in which the script fails to make a decision. They may occur in different parts of our lemma reconstructor. Firstly, if the compound has to be split without Gertwol's word segmentations, i.e. the word boundaries have to be found based on our corpus words; e.g. *Outdoorübungen* in *Indoor- und Outdoorübungen* (indoor [exercises] and outdoor exercises). Secondly, if the compound word consists of more than two parts, e.g. *Murmeltierkinder* in *Gämsen- und Murmeltierkinder* (chamois [children] and marmot children). Finally, if we have to decide between several patterns in cases where more than one solution is possible, like *abenteuer-/TRUNC und/KON tatenfrohe/ADJA Männer/NN* (adventurous and enthusiastic men)

---

[7] # for a strong boundary, – for a weak boundary, \for a linking element, ~ as suffix marker.

which matches the aforementioned pattern (6). In all these cases, we generate different possible solutions and decide based on frequency of the words in our corpus, which solution is the most likely. Although we generated *Indoorübungen* (indoor exercises), *Gämsenkinder* (chamois children) and *abenteuerfrohe* (adventurous), our script could not decide by the word frequency because these words did not occur in the corpus·and thus had the frequency 0, like all the other generated versions.

Problematic cases are especially proper nouns, incl. different spelling variants (e.g. *Monte-Rosa* vs. *Monterosa*), but also old spellings like *Thal* instead of *Tal* (valley) or *Thee* instead of *Tee* (tea), as these are unknown to Gertwol. An approach which could lead to an improvement is the use of a language model to compute the most probable word segmentation. If French translations of the texts exist, another approach could use the French version of the compound in order to solve the German version; e.g. the French version *refuge Vallot, rempli de neige et de salet* could give a useful hint that *schnee- und dreckgefüllte Vallot-Hütte* (Vallot Hut, filled with snow and mud) should result in *schnee#gefüllt* and not in *schnee-#Hütte*.

False Positives in the Search: Some cases which should not be found, unfortunately have been found. If the PoS tag accidentally matches a pattern and additionally a wrong "solution" is found by mistake, it may lead to incorrect changes of the lemmas. E.g. ... *sieht- und merkbar* ... (... sees and noticeable ...) lead to the "correction" *sieht#bar*. The cause of such problems is usually an incorrectly annotated *TRUNC* tag, hence a tagger error.

False Negatives in the Search: There are still cases which our program does not find. Not only due to the lack of coverage in the patterns, but also because of incorrect PoS tags in the corpus, like *bzw.* as *KOUS* instead of *KON* in *Wald- bzw. die Baumgrenze* (forest [line] resp. the tree line), *als* in *Sport- und als Naturschutzverein* (sports [club] and as nature conservation club) as comparative particle *KOKOM* instead of preposition *APPR*, or *haltlose* in *rat- und haltlose Nihilisten* (help[less] and anchorless nihilists) as verb instead of adjective *ADJA*. Furthermore *False Negatives* also occur because of tokenising or other errors, like in *Berg- u.a. Sportarten* (mountain [sport] and other sport) where *u.a.* has been tokenised as one word and annotated with the PoS tag *ADV*. In order to improve the results, the function could be extended by some more patterns.

Very difficult to handle are patterns in which clauses are squeezed in between the truncated word and the full compound. Sometimes the clause is enclosed in parentheses like in *Beich- oder (laut der Karte) Birch-grat* (Beich [ridge] or (according to the map) Birch ridge). However, in complex noun phrases there is no punctuation to indicate that there is an additional clause in between, e.g. *Knospen- oder für einjährige Gewächse wirklich eine neue Pflanzenwelt* (bud [world] or for one-year-old plants really a new plant world).

Hyphen: As compounds often contain hyphens, like e.g. *Monte-Rosa-Gruppe* (Monte Rosa Massif), our lemma reconstructor handles those cases separately. This means the full compound is segmented at the last hyphen, and the last part is attached to the truncated word, which usually results in the correct completion: *Trango- und Biale-Gruppe* (Trango [Massif] and Biale Massif) becomes *Trango#-Gruppe*. However, it leads to mistakes if a word (possibly due to OCR errors) contains incorrect hyphens. This is what happened with: *Kohlenhydrat- und Kalo-rienzufuhr* (carbohydrate [supply] and calory supply), which resulted in *Kohlenhydrat#-rienzufuhr*.

Morphology: Sometimes we incorrectly select inflected forms as lemma if the word has been chosen as the lemma of the elliptical compound. Normally, the lemma provided by Gertwol is stored as the lemma, so that it contains word segmentation symbols. If Gertwol does not know the word but there is a lemma specified in our corpus file, the latter is saved. If there is no lemma in the corpus file either, the actual word is stored as lemma, which sometimes results in inflected forms rather than base forms.

6   Evaluation

We used the 53 German volumes (1957 - 2009) of the Text+Berg corpus' *Release_145_v02* for the evaluation. In total, 11'292 patterns were found. In 1379 cases, i.e. 12.2% of the found patterns, our lemma reconstructor ran into decision problems. As table 1 shows, the simplest pattern is at the same time the most frequent with 7627 occurrences, which are 67.5% of the total cases found. However, this is not surprising concerning the frequency of noun compounds in German. The TIGER corpus also confirms this number with 996 cases of the total of 1362, which is 73.1%.

To determine the accuracy of our lemma reconstructor, we made two different evaluations with small random samples of the Swiss Alpine Club yearbooks from 1957 to 2009. On the one hand we evaluated the pattern search (section 4.1) and on the other hand we examined whether the found cases were handled properly (section 4.2). That is, we distinguished between the evaluations of the search and of the corrections.

For the evaluation of the **search** we applied the usual definitions of *precision* ($\frac{TP}{TP+FP}$), *recall* ($\frac{TP}{TP+FN}$) and *f-measure* ($2*\frac{P*R}{P+R}$). For the results of the **correction** we ignored *true negatives* and *false positives* because they were not relevant. In that case, only the ratio of $\frac{corrected\ properly}{correctly\ identified\ as\ TRUNC}$ (*precision*) respectively $\frac{corrected\ properly}{all\ correct\ TRUNC}$ (*recall*) were important. This means that cases which should not be found were ignored, as they were evaluated in the evaluation of the search.

Table 1. Frequencies of the found patterns

Freq.	Pattern
7627	**(9) TRUNC + KON + NN**
935	(11) WORD + KON + -WORD
522	(6) TRUNC + KON + ADJA/CARD/ADV + NN/NE
488	(2) TRUNC + $, + TRUNC + KON + NN/NE/ADJA
436	(5) TRUNC + KON + ADJ/ADV/VVFIN/VVPP/CARD/APPR + !NN/NE
306	(7) TRUNC + KON + ART/APPR/APPRART + NN/NE
96	(10) TRUNC + APPRART + NN/NE/ADJA
69	(1) TRUNC + $, + TRUNC + $, + TRUNC + KON + NN/NE/ADJA
64	(3) TRUNC + KON + -WORD
33	(8) TRUNC + KON + ART/CARD/ADV/APPR + ADJA + NN/NE
29	(4) TRUNC + KON + APPR + ART + NN/NE

6.1 Evaluation of the Search

This evaluation was carried out with 106 random sentences, two per book. Additionally we made sure that *TRUNC* tokens occur in each of these sentences, hence the 106 sentences contain 127 *TRUNC* tokens. 65 of which were correctly found by our lemma reconstructor, 56 were correctly not found. Five were not found but should have been found, one was incorrectly found. 49 of the 65 *True Positives* were corrected properly, six changed wrongly and ten were not changed due to decision problems. Furthermore, 50 lemmas were analysed by Gertwol, 15 were segmented with the help of word frequencies.

This means the search evaluation resulted in 98.5% precision, 92.9% recall which corresponds to an f-measure of 95.6%.

6.2 Evaluation of the Corrections

For this experiment, we took 100 random sentences each containing correctly annotated *TRUNC* tokens. In these 100 sentences, 106 *TRUNC* tokens occurred. 100 of them were found by our lemma reconstructor, 89 of which were corrected properly, six changed wrongly and five could not be changed due to decision problems. Six *TRUNC* patterns were not found. Furthermore, 79 lemmas were analysed by Gertwol, 21 were segmented with the help of frequencies. Thus our correction evaluation showed a precision of 89%, a recall of 84%, and an f-measure of 86.4%.

7 Conclusion

We have presented an approach to reconstruct lemmas for truncated words. Our lemma reconstructor analyses the Text+Berg corpus and corrects the lemmas of elliptical coordinated compounds. It looks for specific patterns at the PoS tag level, analyses and splits the compounds and generates solutions depending on the pattern. The splitting of the compounds is supported by a list of words containing word segmentation symbols created by the Gertwol system. The decision between multiple solutions is based on word frequencies which we derived from the corpus itself.

The evaluation of the search reached an *f-measure* of 95.6%, and that of the correction an *f-measure* of 86.4%. Concerning the frequencies of the found patterns, it is remarkable that the simplest construction, namely *TRUNC + KON + NN*, accounts for 67.5% of all matches. The amount of cases with decision problems is 12.2% (1379 of 11'929 found patterns). This shows that the reduction of these cases offers a large potential for improvement, which is planned for future work. This could be achieved, for example, by an additional corpus or by using a language model to compute the most probable word segment which completes the truncated compound. Furthermore, an extension of the search patterns should be taken into consideration to improve the coverage.

Our lemma reconstruction program is open-source, and we are happy to share it with interested parties.

Acknowledgements. The first author would like to thank Paul Buitelaar and colleagues at DERI, Galway for support and valuable comments during a research stay where part of this work was carried out.

References

Brants, S., Dipper, S., Hansen, S., Lezius, W., Smith, G.: The TIGER treebank. In: Proceedings of the Workshop on Treebanks and Linguistic Theories, Sozopol (2002)

Bredel, U.: Die Interpunktion des Deutschen. Niemeyer Verlag (2008)

Eisenberg, P.: Grundriß der deutschen Grammatik. Der Satz. Metzler Verlag, Stuttgart, 2. überarbeitete und aktualisierte edition (2004)

Gustafson-Capková, S., Hartmann, B.: Manual of the Stockholm Umeå corpus version 2.0. description of the content of the SUC 2.0 distribution, including the unfinished documentation by Gunnel Källgren. Technical report, Stockholm University (2006)

Holz, F., Biemann, C.: Unsupervised and knowledge-free learning of compound splits and periphrases. In: Gelbukh, A. (ed.) CICLing 2008. LNCS, vol. 4919, pp. 117–127. Springer, Heidelberg (2008)

Rakić, S.: Some observations on the structure, type frequencies and spelling of English compounds. SKASE Journal of Theoretical Linguistics (2009)

Schmid, H.: Probabilistic part of speech tagging using decision trees. In: Proceedings of International Conference on New Methods in Language Processing, Manchester, UK (1994)

Srinivasan, V.: Punctuation and parsing of real-world texts. In: Sikkel, K., Nijholt, A. (eds.) Proceedings of the Sixth Twente Workshop on Language Technology, Natural Language Parsing, Methods and Formalisms. ACL/SIGPARSE Workshop, Enschede, pp. 163–167 (1993)

Thielen, C., Schiller, A., Teufel, S., Stöckert, C.: Guidelines für das Tagging Deutscher Textkorpora mit STTS. Technical report, IMS und SfS (1999)

Volk, M.: Choosing the right lemma when analysing German nouns. In: Multilinguale Corpora: Codierung, Strukturierung, Analyse. 11. Jahrestagung der GLDV, pp. 304–310. Enigma Corporation, Frankfurt (1999)

Volk, M., Bubenhofer, N., Althaus, A., Bangerter, M., Furrer, L., Ruef, B.: Challenges in building a multilingual alpine heritage corpus. In: Proceedings of LREC, Malta (2010)

Volk, M., Furrer, L., Sennrich, R.: Strategies for reducing and correcting OCR errors. In: Sporleder, C., van den Bosch, A., Zervanou, K. (eds.) Language Technology for Cultural Heritage: Selected Papers from the LaTeCH Workshop Series, Theory and Applications of Natural Language Processing, pp. 3–22. Springer, Berlin (2011)

Error Annotation of the Arabic Learner Corpus

A New Error Tagset

Abdullah Alfaifi[1], Eric Atwell[1], and Ghazi Abuhakema[2]

[1] University of Leeds, Leeds, UK
{scayga,e.s.atwell}@leeds.ac.uk
[2] College of Charleston, SC, USA
AbuhakemaG@cofc.edu

Abstract. This paper introduces a new two-level error tagset, AALETA (Alfaifi Atwell Leeds Error Tagset for Arabic), to be used for annotating the Arabic Learner Corpora (ALC). The new tagset includes six broad classes, subdivided into 37 more specific error types or subcategories. It is easily understood by Arabic corpus error annotators. AALEETA is based on an existing error tagset for Arabic corpora, ARIDA, created by Abuhakema et al. [1], and a number of other error-analysis studies. It was used to annotate texts of the Arabic Learner Corpus [2]. The paper shows the tagset broad classes and types or subcategories and an example of annotation. The understandability of AALETA was measured against that of ARIDA, and the preliminary results showed that AALETA achieved a slightly higher score. Annotators reported that they preferred using AALETA over ARIDA.

Keywords: error, tagset, Arabic, corpus, learner.

1   Introduction

The benefits of learner error annotation are multi-faceted and extend to fields such as Contrastive Interlanguage Analysis (CIA), learner dictionary making, Second Language Acquisition, and designing pedagogical materials. CIA is still one of the most frequently used approaches for analyzing a learner corpus, as it enables researchers to observe a wide range of instances of underuse, overuse, and misuse of various aspects of the learner language at different levels: lexis, discourse and syntax [3]. Analyzing errors will also enable researchers and educators to understand the interlanguage errors caused by L1 transfer, learning strategies and overgeneralization of L1 rules. Secondly, learner corpora were – and still are – used to compile or improve learner dictionary contents, particularly by identifying the most common errors learners make, and then providing dictionary users with more details at the end of relevant entries. These errors are indicated in words, phrases, or language structures, along with the ways in which a word or an expression can be used correctly and incorrectly [3, 4]. Also, error-tagged learner corpora are useful resources to measure the extent to which learners can improve their performance in various aspects of the target language [4, 5]. Compilers of longitudinal learner corpora usually include this goal in

I. Gurevych, C. Biemann, and T. Zesch (Eds.): GSCL 2013, LNCS 8105, pp. 14–22, 2013.

their aims. Examples of these include the LONGDALE project: LONGitudinal DAtabase of Learner English [6], Barcelona Age Factor [7], and the ASU corpus [8]. Finally, analyzing learners' errors may be beneficial for pedagogical purposes such as instructional teaching materials development. It can, for instance, help in developing materials that are more appropriate to learners' proficiency levels and in line with their linguistic strengths and weaknesses.

2 Rationale for Developing a New Tagset for Arabic Learner Corpora

The classification of errors in Arabic texts should take into account the nature of the different aspects of linguistic description (e.g., lexis, morphology, syntax, semantics, etc.), and the tagset used for this classification should be readily understandable. These two principles are applied in a number of error tagsets that are used and are publicly available, such as Dagneaux, Denness [9] – used in the International Corpus of Learner English, Granger [10] – used in the French Interlanguage Database (FRIDA) corpus, Nicholls [11] – used in the Cambridge Learner Corpus, Izumi, Uchimoto [12] – used in the NICT JLE Corpus, and ARIDA [1] – used in the Pilot Arabic Learner Corpus.

Abuhakema et al's ARIDA tagset aforementioned is the sole error tagset specifically created for Arabic learner corpora, and it is based on the French Interlanguage Database FRIDA tagset. This adaptation from a French tagset, however, rendered some classification inconsistency with traditional Arabic linguistics. For example, in traditional Arabic, grammatical and syntactic errors are combined under one category called either grammar or syntax; in the ARIDA tagset, these are two different error categories. We recognize however that ARIDA's classification may prove appropriate to those trained in Romance languages where this distinction exists. Moreover, the ARIDA tagset is a three-layered tagset that include error domains, grammar categories and error categories. With a language as diverse as Arabic, we felt that two layers of tagging might be sufficient, and training annotators can be a less daunting task for the new tagset. While the ARIDA tagset uses three-character tags, the new tagset uses two-character tags. In addition, a number of the categories in the FRIDA-derived tagset have a literal translation into Arabic with no clarification of what they linguistically or practically mean, which renders them vague. Examples include *Adjective Complementation* "متممة الصفة", *Noun Complementation* "متممة الاسم", and *Verb Complementation* "متممة الفعل". Further, most of the morphological categories describe the error place and not the type. The sole exception is *Inflection confusion* " الخلط في التصريف", which describes an essential morphological error in Arabic learner production. In the *Form/spelling* category, Abuhakema lists important error types, like *hamza* "الهمزة" (ء) and *tanwin* "التنوين" (ًٌٍ), but neglects some others, like *tā' mutaṭarrifa* "التاء المتطرفة" (ـة، ـت), *'alif mutaṭarrifa* "الألف المتطرفة" (ـى، ا), *'alif fāriqa* "الألف الفارقة" (ـوا), and *lām Šamsiya* "اللام الشمسية" (الشّـ).

3   Basis of AALETA Development

As a result of the above limitations, we developed another error taxonomy based on ARIDA and other error-analysis studies [13-16]. The reason for relying on the ARIDA tagset is that it includes two comprehensively well-described categories, Style and Punctuation. The other four studies investigate different real types of error in Arabic learner production using the bottom-up method where they analyzed their own samples then extracted the corresponding error-type lists. These studies do not aim to develop an error-type tagset to be used for further projects, such as learner corpora. Nonetheless, their error taxonomies are valid and adaptable since they include significant and comprehensive classes of learner error. Furthermore, we cannot overlook the authenticity of the texts from which these error types are derived; which adds to the validity of their taxonomies. The following is a brief overview:

- Alosaili [13] investigates errors of Arabic learners in their spoken production. His list of errors consists of three main classes: phonological, syntactic, and lexical errors, with sub-types under each domain. Some of these types are included in the tagset proposed in this study, specifically those related to orthography, as they were well-formed and cover clearly significant types.
- Alateeq [14] focuses on semantic errors and extracts a detailed list of them, which is adapted in the proposed tagset. Aside from these semantic errors, the study also lists several phono-orthographical, morphological, and syntactic types of error.
- Alhamad [15] focuses on the writing production of advanced level Arabic learners, and concludes with a list of error categories: phonological, orthographical, morphological, syntactic, and semantic errors. The most comprehensive errors are under orthography and syntax, which are added to the tagset we created.
- Alaqeeli [16] examines learners' written errors in a particular type of sentence: a verbal sentence "الجملة الفعلية". This study, therefore, has a limited number of error types under two categories: morphological and syntactic. However, errors under the morphological category are deemed worthy of inclusion in the tagset suggested, due to their comprehensiveness.

Table 1. Error taxonomies in some Arabic studies

Alosaili	Alateeq	Alhamad	Alaqeeli
أخطاء في الأصوات Phonological errors	أخطاء صوتية إملائية phono-orthographical errors	أخطاء نحوية Syntactic errors	أخطاء نحوية Syntactic errors
أخطاء في تراكيب Syntactic errors	أخطاء صرفية Morphological errors	أخطاء صرفية Morphological errors	أخطاء صرفية Morphological errors
أخطاء في المفردات Lexical errors	أخطاء نحوية Syntactic errors	أخطاء إملائية Orthographic errors	
	أخطاء دلالية Semantic errors	أخطاء صوتية Phonological errors	
		أخطاء دلالية Semantic errors	

4   AALETA Tagset

As described, there was a need to develop an error tagset that can provide users (e.g., researchers of Arabic, teachers, etc.) with easily understood broad classes or categories and comprehensive error types. The suggested taxonomy, AALETA, includes 37 types of error, divided into 6 classes or categories: orthography, morphology, syntax, semantics, style, and punctuation. AALETA has two levels of annotation in order to simplify its use and evaluation at this early stage of development. A third layer can be added later when these two layers have achieved a high percentage of accuracy in their use. Each tag consists of two Arabic characters (with an equivalent tag in English). The first character in each tag indicates the error class or category (Table 2), while the second symbolizes the error type (see the example of morphological error in Table 3). For example, the tag *OH* indicates an *[o]rthographical* error in *[H]amza*.

Table 2. Representing error categories in the tagset

Error Category	Orthography الإملاء	Morphology الصرف	Syntax النحو	Semantics الدلالة	Style الأسلوب	Punctuation علامات الترقيم
First part in the Arabic tags	إ	ص	ن	د	س	ت
First part in the English tags	O	M	X	S	T	P

Table 3. Examples of error types (under the morphological category)

Morphological error الأخطاء الصرفية	Word inflection صيغة الكلمة	Verb tense زمن الفعل	Other morphological errors أخطاء صرفية أخرى
Second part in the Arabic tags	ص	ز	خ
Second part in the English tags	I	T	O

This taxonomy is flexible and is to be modified based on studies, evaluation, or relevant results. In addition, at the end of each category, there is an item named "*Other [...] errors*", which can handle any error(s) that do not yet have match(es).

Table 4. AALET: error taxonomy for Arabic learner corpora

Error Category مجال الخطأ	Error Type نوع الخطأ	A-tag الرمز العربي	E-tag الرمز الإنجليزي
Orthography الإملاء 'al'imlā'	1. *hamza* (ء، أ، إ، ؤ، ئ، ئـ) الهمزة	‹إه›	‹OH›
	2. *tā' mutaṭarrifa* (ة، ت) التاء المتطرفة	‹إة›	‹OT›
	3. *'alif mutaṭarrifa* (ا، ى) الألف المتطرفة	‹إى›	‹OA›
	4. *'alif fāriqa* (كتبوا) الألف الفارقة	‹إت›	‹OW›
	5. *lām Šamsiyya* (الطالب) اللام الشمسية	‹إل›	‹OL›
	6. *tanwin* (ٌٍَ) التنوين	‹إل›	‹ON›
	7. *fasl wa wasl* (Conjunction) الفصل والوصل	‹إو›	‹OF›
	8. Shortening the long vowels تقصير الصوائت الطويلة (ٌٍَ → اوي)	‹إف›	‹OS›
	9. Lengthening the short vowels تطويل الصوائت القصيرة (اوي → ٌٍَ)	‹إق›	‹OG›
	10. Wrong order of word characters الخطأ في ترتيب الحروف داخل الكلمة	‹إط›	‹OC›
	11. Replacement in word character(s) استبدال حرف أو أحرف من الكلمة	‹إس›	‹OR›
	12. Character(s) redundant وجود حرف أو أحرف زائدة	‹إز›	‹OD›
	13. Character(s) missing وجود حرف أو أحرف ناقصة	‹إن›	‹OM›
	14. Other orthographical errors أخطاء إملائية أخرى	‹إخ›	‹OO›
Morphology الصرف 'aṣṣarf	15. Word inflection صيغة الكلمة	‹صص›	‹MI›
	16. Verb tense زمن الفعل	‹صز›	‹MT›
	17. Other morphological errors أخطاء صرفية أخرى	‹صخ›	‹MO›
Syntax النحو 'annaḥw	18. Case/Mood Mark الموقع الإعرابي أو علامة الإعراب	‹نب›	‹XC›
	19. Definiteness التعريف والتنكير	‹نع›	‹XF›
	20. Gender التذكير والتأنيث	‹نذ›	‹XG›
	21. Number (Singular, Dual and plural) العدد (الإفراد والتثنية والجمع)	‹نف›	‹XN›
	22. Word(s) order ترتيب المفردات داخل الجملة	‹نت›	‹XR›
	23. Word(s) redundant وجود كلمة أو كلمات زائدة	‹نز›	‹XT›
	24. Word(s) missing وجود كلمة أو كلمات ناقصة	‹نن›	‹XM›
	25. Other syntactic errors أخطاء نحوية أخرى	‹نخ›	‹XO›
Semantics الدلالة 'addalāla	26. Word selection اختيار الكلمة المناسبة	‹دب›	‹SW›
	27. Phrase selection اختيار العبارة المناسبة	‹دق›	‹SP›
	28. Failure of expression to indicate the intended meaning قصور التعبير عن أداء المعنى المقصود	‹دد›	‹SM›
	29. Wrong context of citation from Quran or Hadith الاستشهاد بالكتاب والسنة في سياق خاطئ	‹دس›	‹SC›
	30. Other semantic errors أخطاء دلالية أخرى	‹دخ›	‹SO›
Style الأسلوب 'al'uslūb	31. Unclear style أسلوب غامض	‹سغ›	‹TU›
	32. Prosaic style أسلوب ركيك	‹سض›	‹TP›
	33. Other stylistic errors أخطاء أسلوبية أخرى	‹سخ›	‹TO›
Punctuation علامات الترقيم 'alāmāt 'at-tarqīm	34. Punctuation confusion الخلط في علامات الترقيم	‹تط›	‹PC›
	35. Punctuation redundant علامة ترقيم زائدة	‹تز›	‹PT›
	36. Punctuation missing علامة ترقيم مفقودة	‹تن›	‹PM›
	37. Other errors in punctuation أخطاء أخرى في علامات الترقيم	‹تخ›	‹PO›

5   Scope of Error Tags

The following example, from the Arabic Learner Corpus[1], includes two errors, ortho-graphical OT: character redundant in اللتي "which" [*'allatī*]) and stylistic TP: prosaic style in أعطيت أنا لك "I gave you" [*'a'ṭaytu 'anā 'anta*]). It demonstrates how these errors can be annotated with the appropriate tags when the error is one morpheme (first error) or more (second error). Beside the error annotation, the example here shows lemmas, part-of-speech, and grammatical function tags, and a method of word segmentation in XML (Extensible Markup Language) format:

```
<err type="OD" errform="اللتي" crrform="التي">
  <w>اللتي
    <t token="اللتي" lemma="التي" pos="NR" fun="VA"></t>
  </w>
</err>
<w>كنت
    <t token="كن" lemma="كان" pos="VP"></t>
    <t token="ت" lemma="ت" pos="RR" fun="NK"></t>
</w>
<w>قد
    <t token="قد" lemma="قد" pos="PB"></t>
</w>
<err type="TP" errform="أعطى أنا لك" crrform="أعطيتك">
  <w>أعطى
    <t token="أعطى" lemma="أعطى" pos="VP"></t>
  </w>
  <w>أنا
    <t token="أنا" lemma="أنا" pos="NP" fun="NV"></t>
  </w>
  <w>لك
    <t token="ل" lemma="ل" pos="PP"></t>
    <t token="ك" lemma="ك" pos="RR" fun="GF"></t>
  </w>
</err>
```

6   Measuring Understandability of AALETA

To measure the understandability of AALETA against the tagset developed by Abuhakema et al. [1], two annotators (indicated by T1 and T2) were asked to find errors in a sample of learner texts (the same sample for each annotator), and to mark these errors with tags using the proposed refined taxonomy. Both annotators have masters' degrees and have taught Arabic as a Foreign Language for several years. However, they have not worked on corpus analysis or been involved in any similar

---

[1] ALC is accessed from: http://www.comp.leeds.ac.uk/scayga/alc

task. This can be an advantage, as it could reveal the extent to which the tagset can be understood and useable by untrained users. The texts were taken from ALC which comprises a collection of texts written by learners of Arabic in Saudi Arabia. The corpus covers two types of students, non-native Arabic speakers (NNAS) learning Arabic as a second language (ASL) for academic purpose (AAP), and native Arabic speaking students (NAS) learning to improve their written Arabic. Both groups are males at pre-university level.

Each annotator had to tag the texts twice, using ARIDA tagset first, and AALETA second. Annotators were asked to add the same tag to each repeated error. The assumption was that both error tagsets were clear enough to both annotators, and that they understood which tag is most appropriate to use. Therefore, the error categories and types of both tagsets (ARIDA and AALETA) were not explained to the annotators. This measurement may be sufficient to check whether a tagset can be independently understood against another tagset, considering that the differences between annotators are sometimes due to the annotator's view of the error type, and not to tagset clarity.

The results show that T1 detected 80 errors, while T2 found 91, and they shared 42 errors; the comparison was performed by calculating matched tags between T1 and T2 in each tagset. When the annotators used the ARIDA tagset, they added the same *error-category* tags to 15 errors (36%) out of 42, and the same *error-type* tags to 14 errors (33%). By using AALETA, the annotators shared the same *error-category* tags on 27 errors (64%), and the same *error-type* tags on 22 errors (52%). Although AALETA achieved a higher score, it is still not perfect, which means that it needs more refinement, and that more tests are still needed using other texts and more annotators.

Determining whether a word/phrase was right or wrong was completely based on the annotator's view. It was very likely that some differences in their decisions, particularly in some categories such as semantics and style, relate to the degree of linguistic knowledge of the annotator. The disagreements might have been minimized if annotators were given texts with errors already identified and were asked to mark the appropriate tag on each error. This method can be used in future experiments to avoid such differences.

Table 5. Annotating comparison between Abuhakema and AALETA error tagsets

Using Abuhakema's tagset		
	Error Category	Error Type
No. of same tags (out of 42)	15	14
Percentage	36%	33%

Using AALETA		
	Error Category	Error Type
No. of same tags (out of 42)	27	22
Percentage	64%	52%

When the annotators were asked "*Which taxonomy was more understandable? And why?*", both selected AALETA because of the logical order of its items, and its comprehensiveness. For the question "*Which of them was quick and easy for annotating? And why?*", they both chose AALETA, as they believe that by using AALETA it is easier to select the proper tag, and that the tags are clearer with no ambiguity or overlap.

7   Conclusions and Further Work

This paper introduces a newly-refined tagset for error annotation developed specifically for tagging Arabic learner corpora, and draws on ARIDA and other error classification studies. While ARIDA has its own advantages, we believe that it can be improved in ways that make the annotators' task less daunting. The tagset was used for tagging texts taken from the ALC at two levels: broad classes and error types. An example of the tagging process is presented. The understandability of AALETA was measured against the ARIDA tagset. Although AALETA scored higher, further work is still needed to compare the two tagsets in more detail. Also, to minimize differences in classifying errors, texts with errors already marked can be given, where the annotators' task is to identify the error category and type. This test will present more reliable data about the validity level of each tagset. Thus further work in collaboration with specialists in corpus linguistics and Arabists is still needed – to refine AALETA to increase its suitability for use in further Arabic learner corpora as a standard error tagset, and affirm its understandability over ARIDA. To make it comprehensible and offer more information about learners' errors, another layer may need to be developed and assessed in terms of comprehensibility, validity and applicability. Since the texts were written by male students in one country, diversifying those texts to include more learners from both genders and other countries may yield different results and types of errors.

References

1. Abuhakema, G., Feldman, A., Fitzpatrick, E.: ARIDA: An Arabic Interlanguage Database and Its Applications: A Pilot Study. Journal of the National Council of Less Commonly Taught Languages (JNCOLCTL) 7, 161–184 (2009)
2. Alfaifi, A., Atwell E.: المدونات اللغوية لمتعلمي اللغة العربية: نظامٌ لتصنيف وترميز الأخطاء اللغوية (in Arabic). Arabic Learner Corpora (ALC): A Taxonomy of Coding Errors. In: 8th International Computing Conference in Arabic (ICCA 2012), Cairo, Egypt, 26-28 December 2012 (2012)
3. Granger, S.: The International Corpus of Learner English: A New Resource for Foreign Language Learning and Teaching and Second Language Acquisition Research. TESOL Quarterly 37(3), 538–546 (2003)
4. Nesselhauf, N.: Learner Corpora and Their Potential in Language Teaching. In: Sinclair, J. (ed.) How to Use Corpora in Language Teaching, pp. 125–152. Benjamins, Amsterdam (2004)

5. Buttery, P., Caines, A.: Normalising Frequency Counts to Account for 'opportunity of use' in Learner Corpora. In: Tono, Y., Kawaguchi, Y., Minegishi, M. (eds.) Developmental and Crosslinguistic Perspectives in Learner Corpus Research, pp. 187–204. John Benjamins, Amsterdam (2012)

6. Meunier, F., et al.: The LONGDALE (Longitudinal Database of Learner English), [cited 2012, September 14] (2010), http://www.uclouvain.be/en-cecl-longdale.html

7. Diez-Bedmar, M.B.: Written Learner Corpora by Spanish Students of English: an overview. In: Gómez, P.C., Pére, A.S. (eds.) A Survey on Corpus-based Research, Proceedings of the AELINCO Conference, pp. 920–933. Asociación Española de Lingüística del Corpus, Murcia (2009)

8. Hammarberg, B.: Introduction to the ASU Corpus, a Longitudinal Oral and Written Text Corpus of Adult Learners' Swedish with a Corresponding Part from Native Swedes. Stockholm University, Department of Linguistics (2010)

9. Dagneaux, E., et al.: Error tagging manual (1996)

10. Granger, S.: Error-tagged Learner Corpora and CALL: A Promising Synergy. CALICO Journal 20(3), 465–480 (2003)

11. Nicholls, D.: The Cambridge Learner Corpus - error coding and analysis for lexicography and ELT. In: Corpus Linguistics 2003 Conference (CL 2003), Lancaster, UK (2003)

12. Izumi, E., Uchimoto, K., Isahara, H.: Error anotation for corpus of Japanese learner English. In: Sixth International Workshop on Linguistically Interpreted Corpora (LINC 2005), Jeju Island, Korea, October 15 (2005)

13. Alosaili, A.I.: الأخطاء الشائعة في الكلام لدى طلاب اللغة العربية الناطقين بلغات أخرى: دراسة وصفية تحليلية (in Arabic). Common Errors in Speech Production of Non-Native Arabic Learners, Al Imam Mohammad Ibn Saud Islamic University, Riyadh, Saudi Arabia (1985)

14. Alateeq, Z.M.: تحليل الأخطاء الدلالية لدى دارسي اللغة العربية من غير الناطقين بها في مادة التعبير الكتابي (in Arabic). Semantic Errors Analysis of Non-Native Arabic Learners in Writing, Al Imam Mohammad Ibn Saud Islamic University, Riyadh, Saudi Arabia (1992)

15. Alhamad, M.M.: تحليل أخطاء التعبير الكتابي لدى المستوى المتقدم من دارسي العربية غير الناطقين بها في جامعة الملك سعود (in Arabic) Writing Errors Analysis of Advanced-Level Arabic Learners at King Saud University. Al Imam Mohammad Ibn Saud Islamic University, Riyadh, Saudi Arabia (1994)

16. Alaqeeli, A.S., تحليل الأخطاء في بعض أنماط الجملة الفعلية للغة العربية في الأداء الكتابي لدى دارسي المستوى المتقدم (in Arabic)" Error Analysis in Some Verbal Sentence Patterns of Arabic in Writing Production of Advanced-Level Learners, Al Imam Mohammad Ibn Saud Islamic University, Riyadh, Saudi Arabia (1995)

TWORPUS – An Easy-to-Use Tool for the Creation of Tailored Twitter Corpora

Alexander Bazo, Manuel Burghardt, and Christian Wolff

University of Regensburg, Media Informatics Group, 93040 Regensburg, Germany

Abstract. In this paper we present TWORPUS, an easy-to-use tool for the creation of tailored Twitter corpora. TWORPUS allows scholars to create corpora without having to know about the Twitter *Application Programming Interface* (API) and related technical aspects. At the same time our tool complies with Twitter's "rules of the road" on how to use tweet data. Corpora may be composed in various sizes and for specific scenarios, as the TWORPUS' interface provides controls for filtering and gathering customized collections of tweets, which may serve as the basis for subsequent analyses.

Keywords: Twitter API, web corpora, social media corpora, corpus tool, corpus creation.

1 Introduction

Since the days of the linguist and stenographer Friedrich Wilhelm Kaeding, who with the help of hundreds of assistants manually created and analyzed a corpus of approximately 11 million words from 1891-1897 (Kaeding, 1898), *corpus linguistics* has gained a significant boost from the developments in computerization and data processing. The *World Wide Web* (WWW) plays an important role in this development, as it provides an abundance of machine-readable, freely and ubiquitously available texts (Fletcher, 2012). With the rise of social media platforms like *Facebook.com* and *Twitter.com* in the past decade we also have access to large amounts of user generated content, which allows insights into actual (computer-mediated) language samples to empirically analyze linguistic, political or social issues.

1.1 Web Corpora

Although the web provides large scale and easily accessible language data, it has been discussed whether such data can be used as a corpus without concern. Kilgarriff & Grefenstette (2003) conclude that the web can generally be used as a corpus, but that it depends on the context and type of research question whether the web is a good and suitable resource. They also note, that the requirement of *representativeness*[1] is a general problem of any kind of corpus, and that the web

---

[1] An extensive discussion on "representativeness as the holy grail" in the context of web corpora can be found in Leech (2007).

I. Gurevych, C. Biemann, and T. Zesch (Eds.): GSCL 2013, LNAI 8105, pp. 23–34, 2013.
© Springer-Verlag Berlin Heidelberg 2013

as a corpus is only representative of itself. A prominent example for corpora built from the web can be found in *WaCky (Web-as-Corpus kool initiative)*[2], a working group of researchers interested in using the web as a corpus for linguistic studies (Baroni et al., 2009). The authors also provide an extensive review of other web corpus projects in the related work section.

1.2   Social Media Corpora

Social media have become an important source for collecting current language usage data (Beißwenger & Storrer, 2008). As most social media services are still being operated via the World Wide Web, corpora drawn from them may be categorized as a special kind of web-based corpora.[3] At the same time they are an important driving force of language change, as computer-mediated communication typically differs from other communication channels (Androutsopoulos, 2004, Crystal, 2007, ch. 24, Squires, 2010).

For many social media platforms like *Facebook*, *YouTube* or *Twitter*, APIs are available that allow to draw large samples of textual data for corpus creation. Twitter is the most prominent and dominant type of microblogging services, which allows individuals to publish short messages ("tweets") of up to 140 characters ("SMS of the Web") that can be read by others subscribing to the respective Twitter channel. Among the people with the most followers are idols of popular culture like Justin Bieber or Lady Gaga, each of whom have more than 35 million followers on Twitter.[4] "Hashtags" (e.g. #gscl2013) can be used as descriptors within the tweet message, allowing to search for thematically related tweets. Tweets may also be "retweeted", i. e. a user can republish an existing tweet from another user for his own set of followers. In 2012, Twitter had more than 500 million users.[5]

Analysis of tweets has quickly become an interdisciplinary field of research. For Twitter alone, the cross-disciplinary bibliographic database *Web of Knowledge* (WOK) lists 90 entries with a publication time range from 2009 to 2013.[6] Among the studies looking into Twitter data are as diverse research questions as analyzing tweets as electronic words of mouth in an E-Commerce context (Jansen et al., 2009), sentiment detection in tweets (Bae & Lee, 2012) or using Twitter for analyzing dialect variations in American English ("Twitalectology", Russ, 2012).

---

[2] http://wacky.sslmit.unibo.it, accessed April 10, 2013.

[3] This attribution may change as more dedicated social media apps are used on smartphones with no explicit connection to the web.

[4] Cf. the Twitter monitoring platform *twitaholic*, http://twitaholic.com/, accessed April 17, 2013.

[5] For more detailed information on Twitter, see the comprehensive English Wikipedia article on Twitter, which has been marked as a *good article* (cf. http://en.wikipedia.org/wiki/Twitter, accessed April 17, 2013).

[6] ISI WOK search "Topic=(twitter) AND Topic=(microblog*)" (http://www.webofknowledge.com, accessed April 14, 2013).

2   Corpus Creation on Twitter

Although millions of tweets are published every day, it can be challenging for scholars to get access to this data in a way that enables them to build corpora tailored for their specific research questions. Twitter's *Application Programming Interface* (API) for accessing the continuous stream of tweets and the corresponding *"rules of the road"*[7] present some major technical and legal hurdles.

In the following section we will discuss some common approaches for building Twitter corpora, and how Twitter's developer agreement affects them.

2.1   The Twitter APIs

Twitter offers two different APIs that allow the searching and streaming of tweet collections. Developers may retrieve tweets by querying Twitter's *REST API* with different parameters. While it is possible to search for certain hashtags or query terms, this API does not randomize the sample, but rather returns the first tweets that match the query. Also, it is limited in size and timespan.

The *Streaming API* provides direct access to a continuous stream of current tweets. Free of charge access to this stream is limited to a random sample of approximately one percent of all tweets, while access to all tweets is charged and exclusively granted to selected customers. It is also possible to filter the free streaming sample by using certain query parameters, for instance hashtags or user names. Both APIs require the user to implement *GET* or *POST* requests and to interpret the returning result, which can be received in *JSON* or *XML* format. In order to request a larger number of tweets via the API, the user is required to authenticate as a registered Twitter user by means of the *OAuth* mechanism. To integrate the Twitter APIs in existing software tools, developers can make use of a collection of third party bridges and libraries for different programming languages. Scholars not familiar with the described aspects of programming have to rely on existing corpora or given tools to create tailored tweet collections.

2.2   Twitter Corpora

While many linguists have become familiar with utilizing ready-to-use tools to process and query large amounts of language data, only few of them are able to cope with technically more demanding, rather abstract interfaces to such data, for instance Twitter's *REST API* or the *Streaming API*. Therefore it seems obvious that those who are capable of accessing Twitter data via the APIs should create corpora and share them with the rest of the research community. However, since Twitter changed their developer agreement in 2010 it is no longer allowed to redistribute any tweet messages outside the Twitter platform. Consequently,

---

[7] These rules are also known as the Twitter *developer agreement*. They basically describe the terms and policies for using and redistributing data acquired by the Twitter API (cf. https://dev.twitter.com/terms/api-terms, accessed April 10, 2013).

projects like the *Edinburgh Twitter Corpus* (Petrović et al., 2010) with approximately 100 million tweets are no longer available. Recent Twitter corpora[8] may only be distributed as a list of numerical identifiers that allow to reconstruct the tweets and their corresponding metadata (McCreadie et al., 2012). To retrieve the actual text data, researchers have to use existing crawler applications or create their own implementations of the Twitter API.

Besides the technical and legal obstacles that occur during the creation and distribution of Twitter corpora, researchers also have to accept the lack of customization and personalization of such corpora, as most existing corpora are limited to certain languages, time periods or topics. At the same time, filtering generic tweet collections in a way that makes the language data suitable for answering specific research questions, may result in samples that are too small to derive meaningful observations and interpretations.

2.3 Twitter Corpus Tools

Available web-based tools for the creation of Twitter corpora come with various restrictions and may not be tailored to the specific needs of a particular research project. Such tools allow to monitor current tweets that contain certain hashtags (e.g. *TweetTag*[9]) or that match certain words, phrases or queries (e.g. *Tweet-Archivist*[10]). These tools do not maintain an internal database, but rather rely on *Twitter's Search API* to fetch matching tweets on the fly. One major drawback of such tools is that published tweets can only be restored up to a limit of 2.000 tweets or for a time span of 6-7 days[11]. Although both tools support live monitoring of current and upcoming tweets (continuous searches for selected queries each hour), the total corpus size is limited to 50.000 items (TweetArchivist) or a running time of one day (TweetTag). We designed TWORPUS to encounter these restrictions of existing tools.

3 Description of Tworpus

TWORPUS provides an easy-to-use interface that allows scholars to build large, tailored Twitter corpora. Our tool does not require the user to query the Twitter stream via an abstract API and at the same time meets the Twitter developer agreement.

TWORPUS consists of three main parts (cf. Fig. 1), which are described in more detail in the following sections. (1) A server-based crawler component links into Twitter's free streaming API and stores identifiers and corresponding metadata (but not the tweets themselves) in a *MySQL* database. Users can access the database via (2) a web interface (*corpus creation GUI*) and build customized

---

[8] For instance the *TREC Microblogging Corpus* for the years 2011 and 2012 (cf. http://trec.nist.gov/data/tweets/, accessed April 10, 2013).

[9] http://www.tweet-tag.com/index.php, accessed April 10, 2013.

[10] http://www.tweetarchivist.com, accessed April 10, 2013.

[11] https://dev.twitter.com/docs/using-search, accessed April 10, 2013.

corpora in the fashion of a list of tweet identifiers (IDs). The tweet ID sets can
be filtered by metadata parameters such as language or length, which are stored
with each ID in the database. As the reconstruction of large corpora may take
several hours we provide (3) a desktop tool (*corpus extraction tool*) that allows to
import a list of tweet IDs and subsequently builds a full corpus by automatically
downloading the tweets in TXT or XML-format.

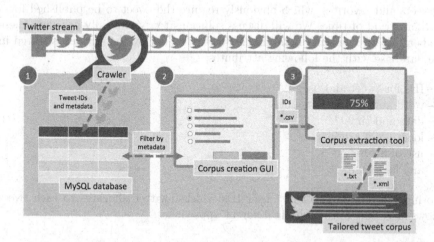

Fig. 1. Overview of the basic workflow and the three main components of TWORPUS

3.1 Dataset, Crawler and Language Detector

Dataset — In the current implementation all data is stored in a single MySQL
database with several relations. Schema and engine are optimized for fast query
processing in large tables. The database and the crawler are hosted on a state
of the art desktop PC running the unix-based operating system *Debian Squeeze*.
An *Apache* web server provides access to the web interface for corpus creation.
Our current test dataset contains approximately eight million tweets. Based on
the crawling speed, we expect the dataset to grow continuously by approximately 600.000 tweets a day. As the free streaming sample is limited to around
1% of *all* tweets, Twitter uses an algorithm to provide a randomized sample.
Unfortunately, Twitter does not provide any information on how the randomization algorithm works[12]. Given the huge amount of overall tweet production
per day,[13] we believe that the sample provided will be large enough for many
relevant research issues.

---

[12] https://dev.twitter.com/docs/faq#6861, accessed April 10, 2013.
[13] *TechCrunch* gives the number of one billion tweets every 2.5 days as of June 2012 (cf.
http://techcrunch.com/2012/07/30/analyst-twitter-passed-500m-users-in-
june-2012-140m-of-them-in-us-jakarta-biggest-tweeting-city/,   accessed
April 17, 2013).

Crawler — The server-based crawler continuously fetches and processes Twitter messages. To connect to the stream we use *Twitter4J* (Yamamoto, 2007), a *Java* bridge to Twitter's APIs. We collect the tweets' actual message content as well as the metadata provided by Twitter (e.g. date and time). In addition, we count the number of *characters* and *words* for each tweet and store this information in the database. With each tweet being crawled in real-time from the stream after its very release, it is not possible to collect information about retweets and favorites, which obviously require the tweet to be published for a certain period of time. We will discuss solutions to dynamically populate these fields afterwards in section 3.3. Each tweet collected by the crawler is stored in the database with the following attributes:

- IDs for tweet and user,
- word and character count,
- date and time[14],
- location of origin,
- use of hashtags,
- and language.

Storing the unique tweet and user IDs which Twitter allocates to each tweet allows later reconstruction via different approaches.

Language Detection — While most attributes are unproblematic and can be stored with clear-cut values, the language information available from Twitter unfortunately is rather ambiguous and unpredictable, as it is based on the settings in the user profile, where each author can define his preferred language. The *preferred language* may however differ from the language that is used in actual tweets, as users tend to write in different languages, or even mix up different languages in the same tweet. To address this problem, we integrated a *language detection library* (Nakatani, 2010) for Java, which uses n-gram frequency profiles to detect the actual language of a tweet. Even though a large number of language profiles is available in this library, several problems occur when using it on Twitter data: As tweets are by nature rather short text fragments, with a maximum length of 140 characters, language detection is challenging, but nevertheless feasible (Gottron & Lipka, 2010). The length restriction for tweets entails heavy use of abbreviations and a generally more telegraphic and fragmentary style of writing. As a result, Bergsma et al. (2012) note that tweets are very heterogeneous in terms of style, and that they are often misspelled and ungrammatical. We have also learned that tweets encoded in non-Latin alphabets may cause additional troubles.

Against this background our recent implementation for detecting the actual language of a tweet can only be considered prototypical and remains an open problem that needs to be addressed in future work. As a result of a series of

---

[14] The timestamp is saved together with a *UTC* offset to allow determination of the local creation time.

detection pretests, the current version of TWORPUS uses the language detection library to identify the following eight languages[15]: Dutch, English, French, German, Italian, Portuguese, Spanish and Turkish.

3.2   Designing a Tailored Twitter Corpus

The corpus creation web interface is realized by means of HTML5 and JavaScript. Users may build a tailored sub-corpus from our dataset (cf. Fig. 2), filtering tweets by the attributes that are stored in the TWORPUS database. The *geolocation* of a tweet or the origin of its author are not implemented as a valid filter criterion. Pretests have shown that the geolocation information that can be gathered from the user profiles is in many cases missing or obviously not realistic, as they do not refer to actual places and cannot be proved to be the actual origin of the tweet. This observation is backed up by Morstatter et al., who found that geolocation information is only available for approximately 3% of the streaming data.

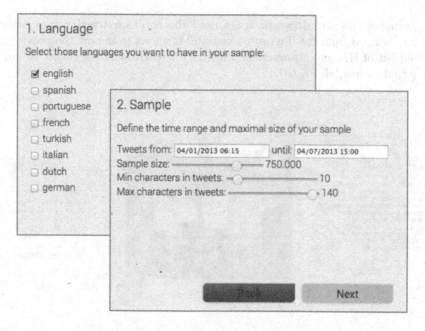

Fig. 2. Corpus creation GUI: Tailored corpora may be designed with regard to language, time and date, sample size and length of tweets

---

[15] The reliability of the language detection is closely connected to the problems described above, including multilingualism and stylistic as well as orthographic aspects of language use in microblogging contexts.

As we aim to provide an easy-to-use tool for the creation of Twitter corpora, the user is guided through the process of corpus creation step by step. Each design step is explained in detail and context sensitive tool tips provide detailed information about the selected filter criteria and possible effects on the sample and its validity. Such information is important as some fields may be misinterpreted (e.g. "language").

Currently, the sample size may range from 10.000 to 1.000.000 tweets. In case the sample size is smaller than the number of available tweets that match the filter criteria, we randomize the sample by using an optimized implementation of SQL's *RANDOM()* function. Once the design parameters for the tweet corpus have been entered, the corpus can be downloaded for further investigation. Complying with Twitter's developer agreement, we only provide a *CSV* file with tweet IDs that allow to build a corpus of actual tweets. For use in later analyses this file also contains information from our database, including word and character counts as well as the detected language. Other metadata will be restored while (re-)building the corpus (cf . section 3.3).

3.3   Building the Corpus

To download the actual tweets, users need the corpus extraction tool, which can be obtained from the TWORPUS website. It allows to import the previously created list of IDs and automatically fetches the corresponding tweets to build the actual corpus (cf. Fig. 3).

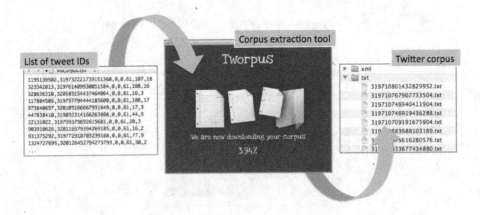

Fig. 3. The corpus extraction tool

The tool generates a folder for the corpus, which contains all tweets in plain text and in XML format. While the plain text files only store the message text of each tweet, the XML file (cf. Fig. 4) contains the respective metadata that is stored in the database as well as up-to-date information fetched from Twitter while downloading. The plain text corpus can be analyzed using "distant reading" (cf. Moretti, 2007) tools such as *Voyant*[16], while the XML-encoded corpus may be investigated with existing query tools such as *XAIRA*[17] or *eXist*[18].

```
<?xml version="1.0" encoding="ISO-8859-1" standalone="yes"?>
<tweet id=308877023269507072>
<user id=579178700>
<screenname>@NewsStocktonCA</screenname>
<fullname>Stockton News</fullname>
</user>
<date>1:49 AM 5 March 13</date>
<retweets>0</retweets>
<favoured>0</favoured>
<text chars=132 words=17 lang=en>Clean air grants offered by local
district: The Yolo-Solano Air Quality Management District Tuesday ...
http://q.gs/3aOXV #stockton</text>
</tweet>
```

Fig. 4. XML representation of downloaded tweet

As downloading the tweets via Twitter's REST API would limit TWORPUS to a maximum speed of 720 tweets per hour[19], we decided to take a different approach: The basic idea is that any published tweet is already available as HTML content on Twitter.com. Each HTML-tweet can be accessed via a unique URL, which contains IDs of the user and his tweet. McCreadie et al. (2012) outline a simple approach for parsing published tweets via the web[20]. By establishing multiple parallel connections on one machine, the speed for downloading the tweets can be improved significantly. In TWORPUS we have implemented 30 parallel threads, which allow to download 10.000 tweets in less than 10 minutes using a broadband Internet connection. Obviously, larger corpora with sample sizes of one million tweets will still take some time to build. Therefore, we implemented a mechanism that automatically resumes a previously paused download process.

McCreadie et al. (2012) also note the dynamic nature of Twitter content as an important aspect that has to be considered: As users can delete tweets or hide them from public access after their release, our dataset may well contain identifiers for tweets that are not available for actual reconstruction via the extraction tool. A list of 10.000 identifiers may result in an actual corpus with a smaller size, as our tool will recognize if a tweet is not available for download

---

[16] http://voyant-tools.org, accessed April 10, 2013.

[17] http://xaira.sourceforge.net/, accessed April 10, 2013.

[18] http://exist-db.org/exist/apps/homepage/index.html, accessed April 10, 2013.

[19] Twitter restricts downloading by GET request to 180 tweets in a rate limited window with a duration of 15 minutes (cf. https://dev.twitter.com/docs/rate-limiting/1.1/limits, accessed April 10, 2013).

[20] As described by McCreadie et al., this technique is also used for downloading the *TREC Microblogging Corpus*.

and exclude it from the corpus. Future releases of TWORPUS will implement a *back channel*, allowing data communication between the download client and the tweet database. When the download client detects unavailable tweets it can query the database for substitutions to accomplish the intended corpus size. The same back channel could be used to gather information about retweets and favorites.

It is important to note, that a corpus extracted from the same list of identifiers at a later point in time, might contain slightly different or modified tweets than a corpus from an earlier extraction, because Twitter users can change or delete their tweets at any time[21]. A corpus extraction in TWORPUS therefore should be treated as a snap-shot of the dynamic and ever-changing *twittersphere*.

4   Outlook

Although in the current implementation TWORPUS is still in its beta testing phase, the web interface for our dataset may be accessed via the corresponding website[22]. The corpus creation tool is also available on the website, and may be downloaded for different operating systems. We strongly encourage other scholars to test and use TWORPUS and are happy to receive feedback on the tool and the dataset. At the same time we intend to work on known limitations and problems of TWORPUS in order to substantially contribute to research with social media corpora.

In the long term, TWORPUS aims at supporting linguistic research. Most of the current literature on tweet analysis focuses on social or political issues of Twitter usage or on *pragmatic* aspects of Twitter language like opinion and sentiment analysis. Little has been published on more traditional linguistic aspects such as lexical, morpho-syntactic or orthographic aspects of Twitter usage. As Twitter does not explicitly mark the actual language used in tweets (which is a vital criterion for linguistic studies), we have implemented a language detection on our own. However, our current language detector still needs to be improved for meeting the special characteristics of grammar, style and spelling in tweets. We are currently planning a crowdsourced study to manually label language, spelling errors, non grammatical terms and other unique characteristics to build language profiles that facilitate language detection on Twitter.

In addition, we are planning to integrate hashtags into the corpus building process. This would enable researchers to generate corpora that do not only match certain language or time span criteria, but also to aggregate tweets for a specific topic.

---

[21] This issue is also described in the TREC microblog track guidelines (cf. https://sites.google.com/site/microblogtrack/2012-guidelines, accessed June 5, 2013).
[22] http://tools.mi.ur.de/tworpus

References

Androutsopoulos, J.K.: Online-Gemeinschaften und Sprachvariation. Soziolinguistische Perspektiven auf Sprache im Internet. Zeitschrift für germanistische Linguistik 31(2), 173–197 (2004)

Bae, Y., Lee, H.: Sentiment analysis of twitter audiences: Measuring the positive or negative influence of popular twitterers. Journal of the American Society for Information Science and Technology 63(12), 2521–2535 (2012)

Beißwenger, M., Storrer, A.: Corpora of Computer-Mediated Communication. In: Lüdeling, A., Kytö, M. (eds.) Corpus Linguistics. An International Handbook, pp. 292–308. Mouton de Gruyter, Berlin (2008)

Baroni, M., Bernardini, S., Ferraresi, A., Zanchetta, E.: The WaCky Wide Web: A Collection of Very Large Linguistically Processed Web-Crawled Corpora. Language Resources and Evaluation 43(3), 209–226 (2009)

Bergsma, S., McNamee, P., Bagdouri, M., Fink, C., Wilson, T.: Language identification for creating language-specific Twitter collections. In: Proceedings of the Second Workshop on Language in Social Media, LSM 2012, pp. 65–74. Association for Computational Linguistics, Montreal (2012)

Crystal, D.: How Language Works. London, Penguin (2007)

Fletcher, W.H.: Corpus analysis of the world wide web. In: Chapelle, C.A. (ed.) Encyclopedia of Applied Linguistics. Wiley-Blackwell (2012)

Gottron, T., Lipka, N.: A Comparison of Language Identification Approaches on Short, Query-Style Texts. In: Gurrin, C., He, Y., Kazai, G., Kruschwitz, U., Little, S., Roelleke, T., Rüger, S., van Rijsbergen, K. (eds.) ECIR 2010. LNCS, vol. 5993, pp. 611–614. Springer, Heidelberg (2010)

Jansen, B.J., Zhang, M., Sobel, K., Chowdury, A.: Twitter power: Tweets as electronic word of mouth. Journal of the American Society for Information Science and Technology 60(11), 2169–2188 (2009)

Kaeding, F.W.: Häufigkeitswörterbuch der deutschen Sprache. Steglitz near Berlin (1898) (self published)

Kilgarriff, A., Grefenstette, G.: Introduction to the Special Issue on the Web as Corpus. Computational Linguistics 29, 333–347 (2003)

Leech, G.: New resources, or just better old ones? The Holy Grail of representativeness. In: Hundt, M., Nesselhauf, N., Biewer, C. (eds.) Corpus Linguistics and the Web, pp. 133–149. Editons Rodopi B.V, Amsterdam (2007)

McCreadie, R., Soboroff, I., Lin, J., Macdonald, C., Ounis, I., McCullough, D.: On building a reusable Twitter corpus. In: Proceedings of the 35th International ACM SIGIR Conference on Research and Development in Information Retrieval, SIGIR 2012, pp. 1113–1114. ACM, New York (2012)

Nakatani, S.: Language Detection Library for Java (website), http://code.google.com/p/language-detection (accessed April 10, 2013)

Moretti, F.: Graphs, Maps, Trees: Abstract Models for a Literary History. London, Verso (2007)

Morstatter, F., Pfeffer, J., Liu, H., Carley, K.M.: Is the Sample Good Enough? Comparing Data from Twitter's Streaming API with Twitter's Firehose. In: ICWSM 2013 (2013), http://www.public.asu.edu/fmorstat/paperpdfs/icwsm2013.pdf (accessed June 3, 2013)

Petrović, S., Osborne, M., Lavrenko, V.: The Edinburgh Twitter corpus. In: Proceedings of the NAACL HLT 2010 Workshop on Computational Linguistics in a World of Social Media, WSA 2010, pp. 25–26. Association for Computational Linguistics, Los Angeles (2010)

Russ, B.: Examining Large-Scale Regional Variation Through Online Geotagged Corpora. Presentation, 2012 Annual Meeting of the American Dialect Society, http://www.briceruss.com/ADStalk.pdf (accessed April 17, 2013)
Squires, L.: Enregistering internet language. Language in Society 39, 457–492 (2010), http://dx.doi.org/10.1017/S0047404510000412 (accessed June 6, 2013)
Yamamoto, Y.: Twitter4J. Java library for the Twitter API (website), http://twitter4j.org (accessed April 10, 2013)

A Joint Inference Architecture for Global Coreference Clustering with Anaphoricity

Thomas Bögel and Anette Frank

Department of Computational Linguistics
Heidelberg University
{boegel,frank}@cl.uni-heidelberg.de

Abstract. We present an architecture for coreference resolution based on joint inference over anaphoricity and coreference, using Markov Logic Networks. Mentions are discriminatively clustered with discourse entities established by an anaphoricity classifier. Our entity-based coreference architecture is realized in a joint inference setting to compensate for erroneous anaphoricity classifications and avoids local coreference misclassifications through global consistency constraints. Defining pairwise coreference features in a global setting achieves an efficient entity-based perspective. With a small feature set we obtain a performance of 63.56% (gold mentions) on the official CoNLL 2012 data set.

1 Introduction

Coreference resolution (CR) is the task of detecting and clustering mentions of discourse entities in a text. Successfully resolving coreferences is an important and challenging step in many NLP tasks. This paper takes a discourse-oriented perspective on CR, by realizing a global, entity-centric approach for mention clustering that exploits dependencies between anaphoricity and coreference. Often, clustering is done after pairwise CR to resolve inconsistencies. We are using a joint approach that performs CR classification and clustering simultaneously, while imposing global consistency constraints. To reduce the complexity of this global clustering approach, we use joint inference with the results of an anaphoricity classifier. Using the anaphoricity classifier, we create anchors for discourse entities and cluster mentions with those entities. Treating both processes in a joint inference architecture ensures that errors of the anaphoricity classifier can be counterbalanced, thus avoiding pipeline effects. Global consistency constraints and exploitation of pairwise features in a global, entity-based coreference formalization enable us to solve the problem efficiently.

Our architecture is based on Markov Logic Networks (Richardson and Domingos, 2006) which combine first-order logic formulas with a probabilistic model. This allows for transparent formalization of the architecture and of the interaction constraints, as well as flexible extensions of the feature set.

This paper is organized as follows: Sec. 2 presents related work. Sec. 3 describes our core architecture for global coreference clustering with anaphoricity. Sec. 4 introduces the components for anaphoricity and coreference determination. Sec. 5 presents our experiments and results and Sec. 6 concludes.

I. Gurevych, C. Biemann, and T. Zesch (Eds.): GSCL 2013, LNAI 8105, pp. 35–46, 2013.
© Springer-Verlag Berlin Heidelberg 2013

2   Modeling Coreference Resolution

Approaches to Coreference Resolution. Machine learning approaches have long addressed CR in a pairwise scenario deciding for each pair of mentions whether they are coreferent or not. A subsequent clustering step then clusters the pairs to entities (Soon et al. (2001), Ng and Cardie (2002)). This is often implemented in a pipeline architecture and thus suffers from error propagation and locality issues, such as violation of transitivity over local CR decisions.

Entity-based approaches, on the other hand, focus on modeling the entity itself by using entity-level features. This reduces the problem of locality as a mention is compared to a complete entity instead of creating an entity by pairwise classification decisions. Defining entity-level features is challenging, as they require appropriate representations for judging similarity between mentions and an entity. Features for entities can, for instance, be defined by aggregating and comparing specific attributes for all mentions in an entity (Wellner et al., 2004).

A further challenge of entity-based approaches is the number of clusters to be examined. Luo et al. (2004) make use of Bell trees to reduce this processing complexity. Martschat et al. (2012) perform global CR using multigraphs, allowing for an entity-based perspective. Finally, ranking approaches are used to find the best antecedent entity for mentions (Rahman & Ng (2011); Denis & Baldridge (2008)) instead of relying on single pairwise classifications.

Joint Inference for Coreference Resolution. To overcome the problem of locality, different aspects of CR can be combined using joint inference techniques. Denis and Baldridge (2009) use Integer Linear Programming to perform joint inference over independent classifier decisions for anaphoricity, pairwise coreference and named entity type. They define global ILP constraints over these classifications to determine a globally optimal solution that respects dependencies of anaphoricity and coreference decisions. In contrast to our approach, pairwise coreference decisions need to be harmonized by explicit transitivity constraints. These could not be fully implemented due to efficiency constraints. Recently, Song et al. (2012) proposed a joint inference formalization using MLN that interfaces local, pairwise resolution and clustering by way of explicit transitivity constraints. We, in contrast, will use anaphoricity as an anchor for entities and perform best-first clustering without stating explicit transitivity constraints.

Poon and Domingos (2008) use MLN to perform unsupervised CR in an entity-based approach that, like ours, implicitly accounts for transitivity. Their formalization does not model interactions with anaphoricity and offers a restricted set of entity-level features. Clustering of mentions is driven by head features, and few semantic type and morphological features are used to assign further mentions to these clusters. Other factors, such as distance, are encoded using a pre-defined prior. Thus their formalization is not truly transparent from a linguistic modeling point of view. As many rules are defined as hard constraints, it is unclear whether the system can be adapted to a supervised scenario, in order to adapt to specific domains, and whether an extended set of features can be integrated using hard constraints or independently determined priors.

2.1   Markov Logic Networks

Markov Logic Networks (MLN) combine two major advantages desirable for NLP applications: (1) being able to model uncertainty efficiently (using probabilistic graphical models) and (2) expressing and combining various sources of knowledge (by first-order logic) (Richardson and Domingos, 2006).

A MLN consists of a predicate logic knowledge base (formulas) with a weight attached to each formula. The weight of a formula can be regarded as the cost of violating it. A high accumulated cost over all formulas leads to a less probable possible world and vice versa. More formally, the probability of a specific state of a world x can be described in a log-linear model as follows:

$$P(X = x) = \frac{1}{Z} \exp(\sum_i w_i n_i(x)) \tag{1}$$

The weight w_i of each formula is multiplied by the number of true groundings of the formula in the specified world $(n_i(x))$.

Each MLN instantiates a Markov network: binary nodes in the network correspond to possible groundings of first-order predicates and one feature (edge) for each formula with the corresponding weight. Each set of ground atoms from a knowledge base thus leads to a specific ground Markov network.

Learning. Learning weights discriminatively for a specific predicate is a problem of maximizing the conditional probability of a query given a possible world: $P(Y = y|X = x) = \frac{1}{Z_x} \exp \sum_i w_i n_i(x, y)$, where x are evidence predicates and y the query predicate. $n_i(x, y)$ counts the number of true groundings of the ith formula. Weights are learned to maximize the conditional probability. Intuitively, the number of true groundings in the data is compared to the expected number of true groundings. With a gradient descent method, this problem can be solved efficiently (Lowd and Domingos, 2007)

Inference. MAP inference aims at finding the most probable state y (query predicate) for given evidence x ($\arg\max_y P(y|x)$), which can be expressed as $\sum_i w_i n_i(x, y)$. Satisfiability solvers solve this NP-hard problem efficiently.

We use Tuffy (Niu et al., 2011) as an inference framework for MLN due to its efficiency during inference. Tuffy implements the Newton Diagonal method for learning weights and MaxWalkSAT for inference.

3   Global CR Clustering with Anaphoricity Using MLN

Our approach aims at modeling CR using a *global, entity-based approach* that will be complemented by a *discourse-based perspective*, by using *discourse-new*, or *anaphoricity* information to guide the entity clustering process. This architecture will be formulated as a joint inference problem using MLN.

```
1 // predicate declarations    6 // discourse-new mentions initiate an entity
2 *cand_ent(MEN,ENT)            7 cand_ent(m,e) , anaph(m,0)  ⇒ m_in_e(m,e)
3 m_in_e(MEN,ENT!)             8 cand_ent(m,e) , anaph(m,1)  ⇒ !m_in_e(m,e)
4 *anaph(MEN,BOOLEAN)
5 *feat(MEN,FEAT)              9 // rule schema for pairwise CR clustering
                              10 m_in_e(m_e,e), !m_in_e(m,e) , m != m_e,
                                 feat(m_e,m,val) ⇒ m_in_e(m,e)
                              11 // constraint: resolve all mentions
                              12 cand_ent(m,e) ⇒ m_in_e(m,e_x) .
```

Fig. 1. Core predicate declarations and rules. , indicates conjunction, ⇒ a conditional and ! negation. * specifies a closed-world assumption and . defines a hard constraint.

We realize a global, entity-based CR system by clustering all mentions into entities and choosing a partition of entities that simultaneously satisfies all coreference features. Such an approach circumvents the problem of local classifications. Implementing this in a brute-force manner is computationally prohibitive, as it requires considering the probability of each possible clustering of mentions. As $\mathcal{P}(M) = 2^n$ for a set M of n mentions, the number of possibilities to compute increases exponentially with the number of mentions. Luo et al. (2004) introduce a bell tree to efficiently manage multiple clusterings of mentions using a beam search, but pruning bears the risk of removing globally good results.

Instead of considering all possible clusterings, we will use anaphoricity as a guide to establish discourse entities for subsequent mention clustering. Mentions that are determined by the anaphoricity classifier to be *non-anaphoric*, or *discourse-new*, will be considered as an 'anchor' for a new discourse entity that serves as a reference point for discriminative clustering of mentions classified as *anaphoric*. Combining anaphoricity and CR in a joint inference architecture with soft constraints allows us to compensate for errors in either module. In contrast to Denis and Baldridge (2009), where pairwise classifications are supported by anaphoricity, our clustering approach focuses on entities from the beginning.

We use Markov Logic Networks, as they offer a flexible and transparent framework that can be easily extended to incorporate additional knowledge.

3.1 Global Architecture and MLN Formalization

Our MLN formalization consists of four main components. We first give a brief overview of these components and their interplay and then turn to discuss them in more detail. Fig. 1 states relevant predicate declarations and rules.

1. **Initialization.** *All* mentions m are declared as potential anchors for an entity e, through a predicate cand_ent(MEN,ENT) (l. 2) and appropriately defined knowledge base entries.
2. **Establishing Entities through Anaphoricity.** Mentions m classified as discourse-new by an external classifier instantiate a unique discourse entity e (the one associated with m through cand_ent(m,e)), through the coreference

indicating predicate m_in_e(m,e). (l. 3, 7). Mentions classified as anaphoric do *not* initiate a discourse entity (l. 8).

3. **Global Coreference Resolution.** The rule schema in line 10 applies to *all* mentions m that are not (yet) clustered with an entity e, and evaluates the strength of individual CR features *feat* (l. 5) holding between m and any distinct mention m_e clustered with e.This implements an entity-based CR strategy while using mention-level features. The rule quantifies over all entities e that were established by a discourse-new mention in **2.** (l. 7).

4. **Two Constraints** ensure consistency and constrain the clustering of mentions to entities:

   a. Uniqueness: *Each mention is assigned to exactly one entity.* This constraint (defined through m_in_e(MEN,ENT!), l. 3) implies *disjointness of discourse entities*. It considerably reduces the combinatorics of clusterings, since anchor mentions m_e of entities e need not be considered by the coreference clustering rules in **3**: a mention m_e that is already clustered with an entity e cannot be clustered with a disjoint entity $e_x \neq e$.

   b. Resolve All Mentions. This constraint (l. 12) enforces that each mention is clustered with an entity: the entity of which it is an anchor $(e = e_x)$ or some other entity $(e \neq e_x)$. When resolving gold mentions, we define it as a hard constraint. It can be relaxed when dealing with system mentions.

   Transitivity. Unlike Denis&Baldridge (2009) and Song et al. (2012) we do not encode explicit transitivity constraints. Violations of transitivity are excluded by **Uniqueness**. Since clustering is entity-centric (through m_in_e(m,e)), the uniqueness constraint implies *disjointness* of entities, and thus, discriminative clustering of mentions to entities.[1]

3.2 Using Anaphoricity to Establish Entities

Anaphoricity determines whether a mention is anaphoric (i.e., refers to a previously mentioned entity) or non-anaphoric and thus introduces a new discourse entity. Mentions classified as discourse-new can be used as 'seeds' for establishing entities for subsequent (discriminative) clustering of anaphoric mentions to the created discourse entities. We exploit linguistic knowledge about anaphoricity to establish the set of discourse entities in a given text, and thereby considerably reduce the space of possibilities to be investigated for entity clustering.

Anaphoricity of mentions is determined by an external anaphoricity classifier (AC) (see Sec. 4.1) that provides binary anaphoricity labels (l. 7, 8). Rule 7 receives high weight with mentions classified as *discourse-new (non-anaphoric)*, and instantiates a discourse entity, of which it is typically the first member. Rule 8 applies to the complementary case, with mentions classified as *anaphoric*. Here, the rule states that m does not instantiate a discourse entity.

Reducing Complexity. Our approach provides an initial partitioning of mentions into a set of anchored discourse entities. Since *uniqueness* forces each mention

---

[1] A sufficiently high weight of the AC rule avoids clustering all mentions into a single entity, as it triggers the creation of a sufficient number of entities.

to be contained in a single entity, mentions serving as anchors for an entity cannot be clustered with other entities. This knowledge about disjointness severely reduces the combinatorics of coreference-based clustering of mentions to entities.

Avoiding Pipeline Effects. While our approach reduces the number of clusterings to be examined, errors of the AC may severely harm CR performance (Denis and Baldridge, 2009). To avoid pipeline effects, the rules that create discourse entities are weighted to counterbalance errors of the AC: We allow for *each* mention to instantiate a new entity with a learned weight that depends on the AC result. At present, we use two rules (l. 7, 8) to represent the binary AC predictions.

3.3   Global Coreference Resolution

There are different ways for using CR features in an entity-based architecture: *cluster- or entity-level features*, i.e., features defined between a mention and an entity, or *mention-level features*, as typically used in *mention-pair* models. As the design of entity-level features is challenging for some feature types (e.g. distance measures) and computing feature values for all possible entity clusterings is computationally expensive, we make use of *mention-level features* that are evaluated in an *entity-centric, discriminative ranking approach* by exploiting universal quantification of variables in MLN formulae (Fig. 1, l. 10).

A mention m that is to be classified will be evaluated for coreference with every entity e by comparing it to all mentions in the cluster and accumulating the evidence for a mention to be in a specific cluster using the rule weights for individual pairwise coreference features. In this way, we compare a mention to *all* other mentions of the entity using pairwise, mention-level features.

In MLN, this intuition is captured by exploiting implicitly universally quantified variables. I.e., for each feature $feat$ and possible value val of $feat$, we employ the rule schema from Fig. 1, l. 10: For all mentions m_e in an entity e: if we observe a feature value for the mention pair (m_e, m), m should also be assigned to e (with rule weight w_{feat}). For features encoding negative evidence, the learned rule weight is negative. According to equation (1), the weight for all instantiated formulas are accumulated. The final weight for assigning mention m to some entity e is thus obtained by comparing m to *all* mentions m_e in cluster e, and we assign m to the entity that receives the highest overall score.

3.4   Joint Inference

We now discuss how the components for anaphoricity and coreference interact to allow for joint inference and ensure consistency for the entire inference process.

Anaphoricity and Coreference. Anaphoricity information is used to provide seeds for entity clusterings to which anaphoric mentions are attached. Misclassifications of the AC need to be counterbalanced by coreference features.

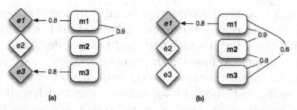

Fig. 2. Joint inference: anaphoricity and coreference

Fig. 2 demonstrates a relevant case: We assume two possible inference results (a) and (b) for a hypothetical document that contains three mentions ($\{m_1, m_2, m_3\}$). Each mention could instantiate a new entity ($\{e_1, e_2, e_3\}$). The AC predicts that mentions m_1 and m_3 are discourse-new which results in a high weight (0.8, in this example) for the mentions to instantiate a new entity (Fig. 1, l. 7). We assume that m_3 is *erroneously* classified as discourse-new. Solution (a) accepts the AC prediction and clusters mentions m_1 and m_2 with e_1 and m_3 as a new entity e_3, whereas solution (b) rejects the AC prediction for m_3.

In addition to anaphoricity, there is an accumulated weight due to all *coreference* indicators between each pair of mentions. For solution (a), the accumulated scores yield an overall weight of $0.8 + 0.8 + 0.9 = 2.7$. Solution (b) rejects the result of the AC and instead attaches m_3 to entity e_1. The overall score of $0.8 + 0.8 + 0.9 + 0.6 = 3.3$ exceeds solution (a). Thus, strong coreference features can indeed override errors of anaphoricity classification.

Conversely, features that contribute strong negative evidence for coreference can help to compensate for discourse-new mentions that are wrongly predicted to be anaphoric. If a mention does not fit with any cluster, it is likely to be non-anaphoric and to instantiate a new entity. Indeed, given strong negative evidence for coreference with any entity, we might obtain a globally optimal solution that establishes an entity using the fallback rule of line 12.

Modeling the mutual interdependence of coreference and anaphoricity decisions using joint inference offers a great advantage over pipeline architectures.

Interaction between Coreference Features. The proposed architecture also handles interactions and contradictions between coreference features. Each mention is evaluated by multiple coreference rules, each one defining different features and values, with different rule weights attached to them. As all rule weights for all features are accumulated (cf. equation (1)), positive or negative evidence for clustering a mention with an entity, are balanced against each other.

4   Anaphoricity and Coreference Features

4.1   Anaphoricity Classifier

We follow the evaluations in Poesio et al. (2005) and selected the most promising features from previous works. We optimized classification performance on the development set of the CoNLL 2012 dataset and chose the Random Forest classifier implemented in Weka (Hall et al., 2009) which yielded highest performance.

For each mention to be classified we determine its mention type, whether it occurs in the first sentence or is the first occurrence of the head or surface form of the mention. We check for pre- and post-modification, definiteness, superlative forms and the grammatical function. In addition to these classical features, we use 8 measures to capture a raise in term frequency and tf-idf after the first mention of an entity that also holds for partial string matches (Ritz, 2010).

Training the anaphoricity classifier on the complete training portion of the CoNLL 2012 data and evaluating it on the development set yields an accuracy of 86.38% (Prec.: 86.5%, Rec.: 86.4%).

4.2   Coreference Features

We selected and implemented 17 features for coreference resolution from (Bengtson and Roth, 2008) and used them to test our architecture.[2] For each feature and possible feature value, we add a dedicated rule. For continuous features (e.g. distances), we first learn weights for each possible feature value individually and subsume values with a similar weight to obtain plausible feature ranges. We re-estimate weights for the obtained feature ranges.

As a fallback, we add a feature that attaches mentions to the *nearest potential antecedent* if there is not enough evidence for coreference. This avoids promoting unbound anaphoric mentions to independent discourse entities.

5   Feature Selection and Experiments

5.1   Experimental Setting

We use CoNLL 2012 Shared Task data (Pradhan et al., 2012) for all experiments and evaluate on the official test set. As our aim in this work is to develop a core baseline architecture as a proof of concept, we focus on *gold standard mentions*. Future work will extend our architecture to include system mentions.

We apply the five evaluation metrics used in the CoNLL 2012 Shared Task: MUC , B^3, CEAF with both the entity- and the mention-based similarity metric and BLANC. The arithmetic mean of all five F_1 scores is used for feature selection and presentation of results.

5.2   Weight Learning and Feature Selection

Despite using pairwise rules, learning weights for many or all rules simultaneously is still computationally expensive: each predicate that is added to the rule file adds a node and edges for each possible grounding of the formula to the Markov

---

[2] Surface features (*HeadMatch, StringMatch, Alias, StringKernelSim.*); Syntactic (*Appositive, Predicative, BindingConstraints, HobbsDistance*); Semantic (*Synonymy, Antonymy, SemanticDistance*); Agreement (*Gend-/Num-/Semantic-Agr*); Distance (*Token-/Mention-/Sent-Dist*).

Table 1. Performance impact of features in additive feature selection: difference relative to last iteration ($\Delta avg(F)$) and absolute performance ($avg(F)$)

HeadMatch	-	59.82
Added rule	$\Delta avg(F)$	$avg(F)$
STRING KERNEL SIMILARITY	+0.56	60.38
GENDER AGREEMENT	+0.20	60.58
NEAREST ANTECEDENT	+0.04	60.62
HOBBS DISTANCE	+1.68	62.30
STRING MATCH	+1.64	63.94
REFLEXIVE MATCH	+1.35	65.29
SEMANTIC DISTANCE	+0.00	**65.29**

network. Dependencies between features in particular result in an exponential growth of the Markov network.

To reduce the size of the resulting Markov network and speed up the learning process, we learn weights for different rules (i.e. coreference feature values) individually. That is, we make a simplifying independence assumption *for all rules* so that we can learn rule weights individually: for independent features, each rule only contains one feature value. Nevertheless, we add the mention type to restrict rules to mentions for which a feature is appropriate. All weights are learned on 50 randomly sampled documents of the CoNLL 2012 training set, containing 2279 mentions in 513 entities.

We extracted features based on the provided automatic annotations and used the output of the AC during training, in order to ensure that the influence of erroneous anaphoricity annotation is learned and counterbalanced.

Additive Feature Selection. To determine an optimal feature set, we conduct greedy forward selection in combination with step-wise backward deletion. We start with a rule set without any coreference features (l. 1-8, 12 in Fig. 1) and add individual coreference rules one at a time to determine which rule yields highest overall performance gain (CoNLL score). This rule is then added to the rule set and the process is repeated by adding further rules until no further improvements are observed. After this process, we perform step-wise backward deletion: at each step, we eliminate one feature. If performance increases after deletion of a single feature, the feature is removed and we continue with forward selection again. This combination of forward selection and backward deletion is repeated until no improvements are observed.

Table 1 lists the features that were selected during the described process, and their contribution to overall performance on the development set.

5.3 Experiments and Results

Evaluation Setup. For final evaluation, we measure CR performance with the selected feature set (Table 1) on the test set using gold mentions and automatically created linguistic annotations.

Table 2. Evaluation results on the test set for four scenarios with optimized features

Evaluation Scenario	(I) AC_{auto} learned weight		(II) AC_{auto} constraint	(III) AC_{gold} learned weight		(IV) AC_{gold} constraint	
documents	*all*	*non-split*	*all*	*all*	*non-split*	*all*	*non-split*
	Prec Rec F_1	F_1	F_1	F_1	F_1	F_1	F_1
MUC	72.55 76.14 74.30	76.31	69.80	75.10	77.08	75.02	77.13
B^3	61.03 67.05 63.90	64.28	58.28	64.20	65.71	64.98	67.31
CEAF-M	64.14 86.54 66.27	66.42	53.26	66.58	66.99	67.47	69.26
CEAF-E	55.47 62.92 58.96	59.92	47.55	59.21	62.93	60.19	62.14
BLANC	52.71 56.14 54.37	60.76	53.21	54.64	62.28	55.00	63.63
CoNLL	**63.56**	65.54	56.42	63.95	67.00	64.53	67.90

Next to the full architecture with joint inference over anaphoricity (AC) and CR classifications (**I**) we evaluate further system variants to highlight the impact of joint inference and the individual anaphoricity and coreference sub modules (cf. Table 2): (**II**) highlights the effect of joint inference against a pipeline of *automatic* AC predictions and a *hard constraint* for establishing entities;[3] (**III**) and (**IV**) use oracle (*gold*) AC results with *learned weights* vs. *hard constraints* for the creation of entities. They illustrate the upper bound of the system's current AC integration and CR clustering performance.

In the CoNLL data, long documents are split into multiple pieces which artificially creates new entities at each break. As our anaphoricity-driven architecture is heavily influenced by such noise, we additionally evaluate most scenarios on the subset of documents that are not affected by such splits (*non-split*).

Results. Our final evaluation results are given in (**I**). With **63.56%**, the F_1-score of our model lies within the range of published results for the CoNLL 2012 Shared Task with gold mentions, where scores range from 51.40% to 77.22%.[4] A pipeline architecture (**II**) suffers from a drop of around 7 percentage points. This clearly shows that our joint inference architecture is effective in counterbalancing AC errors. We note a strong effect for both CEAF metrics.

Scenarios (**III**) and (**IV**) mark upper bounds for our approach regarding AC integration and CR performance. As our AC classifier scores at 81.6/84.7/83.12 $P/R/F_1$ on the test set, the small performance difference (0.39 points F_1) between automatic (**I**) and gold AC using learned rule weights (**III**) shows that the learned rule weights are well set. The results using an AC oracle with hard constraints for entity creation shows small differences, too. This points to a weakness of the current model regarding the performance of CR features.

For all models we observe clear performance increases for non-split documents, which avoids artificial noise that the AC is unable to detect.

---

[3] We simulate a pipeline by marking both anaphoricity rules as hard constraints.
[4] http://conll.cemantix.org/2012/

5.4 Error Analysis and Discussion

Errors involving anaphoricity. Joint inference over anaphoricity and coreference is crucial to our architecture. We thus measured how well joint inference counterbalances **(1)** *false positives (FP)* (mentions erroneously classified as anaphoric) and **(2)** *false negatives (FN)* (anaphoric mentions classified as non-anaphoric).

On the test set, the anaphoricity classifier yields a precision of 81.6% and a recall of 84.7%. In setting **I**, 32% of the FPs and 68% of the FNs are corrected. Thus, errors introduced by the AC erroneously classifying a mention as anaphoric are harder to resolve. Inspection of errors reveals that 52% of the mentions that are not corrected in our joint inference scenario (**I**) are pronouns. Most of these erroneously classified pronouns behave as discourse-new in the gold standard due to the fact that long documents are split into pieces in the CoNLL data set, which artificially creates new entities. As our AC is based on linguistic features, it is deemed to misclassify these mentions as anaphoric. Our approach is especially sensitive to this artificial noise as it depends on correct anaphoricity information. If we remove split files, 43% of FPs and 65% of FNs are corrected.

For FNs, stronger CR indicators are needed such that mentions could be attached to other entities despite being classified as discourse-new.

Discussion and Future Extensions. Our evaluation clearly shows that joint inference over anaphoricity is well designed, so that CR information can counterbalance classifier mistakes. Further extensions will integrate classifier confidence values, to help the impact of CR features for correcting false negatives.

The small feature set we are currently using shows that the architecture itself plus some strong features result in a strong baseline system. At the same time, the evaluation points to weaknesses of our current CR feature set. This is (partially) due to the quite strong independence assumptions during learning. In current work we perform feature selection using linguistically motivated feature groups and also use larger training sets, using a more efficient MLN engine.

6 Conclusion

In this paper we propose an architecture for CR that uses anaphoricity to establish a set of discourse entities in a text and clusters all anaphoric mentions with these entities. By accumulating weights of pairwise mention-level coreference comparisons we realize discriminative mention clustering while circumventing the problem of defining entity-level features. We use Markov Logic Networks as a framework to perform joint inference over the output of an anaphoricity classifier and pairwise entity-centric coreference decisions, and show that the system is able to correct errors of both anaphoricity and coreference. To our knowledge, this is the first attempt to use anaphoricity to establish discourse entities for discriminative global mention clustering. This is what clearly distinguishes our account from Poon and Domingos (2008), where discourse entities are globally clustered by the heads of mentions and agreement features. Our system achieves good performance using a small feature set. We are currently working with feature combinations on larger training set sizes and integrate system mentions to realize a powerful end-to-end system. First experiments yield promising results.

References

Bengtson, E., Roth, D.: Understanding the value of features for coreference resolution. In: Proceedings of EMNLP 2008, pp. 294–303 (2008)

Denis, P., Baldridge, J.: Specialized models and ranking for coreference resolution. In: Proceedings of EMNLP 2008, pp. 660–669 (2008)

Denis, P., Baldridge, J.: Global joint models for coreference resolution and named entity classification. Procesamiento del Lenguaje Natural 42(1), 87–96 (2009)

Hall, M., Eibe, F., Holmes, G., Pfahringer, B., Reutemann, P., Witten, I.: The WEKA Data Mining Software: An Update. ACM SIGKDD Explorations Newsletter 11(1), 10–18 (2009)

Lowd, D., Domingos, P.: Efficient weight learning for markov logic networks. In: Kok, J.N., Koronacki, J., Lopez de Mantaras, R., Matwin, S., Mladenič, D., Skowron, A. (eds.) PKDD 2007. LNCS (LNAI), vol. 4702, pp. 200–211. Springer, Heidelberg (2007)

Luo, X., Ittycheriah, A., Jing, H.: A mention-synchronous coreference resolution algorithm based on the bell tree. In: Proceedings of ACL 2004 (2004)

Martschat, S., Cai, J., Broscheit, S., Mújdricza-Maydt, E., Strube, M.: A Multigraph Model for Coreference Resolution. In: Proceedings of EMNLP-CoNLL 2012: Shared Task, pp. 100–106 (2012)

Ng, V., Cardie, C.: Improving machine learning approaches to coreference resolution. In: Proceedings of ACL 2002, pp. 104–111 (2002)

Niu, F., Ré, C., Doan, A., Shavlik, J.: Tuffy: scaling up statistical inference in Markov logic networks using an RDBMS. Proceedings of the VLDB Endowment 4(6), 373–384 (2011)

Poesio, M., Alexandrov-Kabadjov, M., Vieria, R., Goulart, R., Uryupina, O.: Does discourse-new detection help definite description resolution? In: Proceedings of IWCS, vol. 6, pp. 236–246 (2005)

Poon, H., Domingos, P.: Joint unsupervised coreference resolution with Markov logic. In: Proceedings of EMNLP 2008, pp. 650–659 (2008)

Pradhan, S., Moschitti, A., Xue, N.: CoNLL-2012 shared task: Modeling multilingual unrestricted coreference in OntoNotes. In: Proceedings of EMNLP-CoNLL: Shared Task, pp. 1–27 (2012)

Rahman, A., Ng, V.: Narrowing the modeling gap: a cluster-ranking approach to coreference resolution. Journal of Artificial Intelligence Research 40(1), 469–521 (2011)

Richardson, M., Domingos, P.: Markov logic networks. Machine Learning 62(1), 107–136 (2006)

Ritz, J.: Using tf-idf-related Measures for Determining the Anaphoricity of Noun Phrases. In: Proceedings of KONVENS 2010, pp. 85–92 (2010)

Song, Y., Jiang, J., Zhao, X., Li, S., Wang, H.: Joint Learning for Coreference Resolution with Markov Logic. In: Proceedings of EMNLP-CoNLL 2012, pp. 1245–1254 (2012)

Soon, W., Ng, H., Li, D.: A machine learning approach to coreference resolution of noun phrases. Computational Linguistics 27(4), 521–544 (2001)

Wellner, B., McCallum, A., Peng, F., Hay, M.: An integrated, conditional model of information extraction and coreference with application to citation matching. In: Proceedings of UAI 2004, pp. 593–601 (2004)

Linguistic and Statistically Derived Features for Cause of Death Prediction from Verbal Autopsy Text

Samuel Danso[1,2], Eric Atwell[1,2], and Owen Johnson[2]

[1] School of Computing, Language Research Group, University of Leeds
[2] Yorkshire Centre for Health Informatics, eHealth Research Group, University of Leeds
{scsod,eric.atwell,owen.johnson}@leeds.ac.uk

Abstract. Automatic Text Classification (ATC) is an emerging technology with economic importance given the unprecedented growth of text data. This paper reports on work in progress to develop methods for predicting Cause of Death from Verbal Autopsy (VA) documents recommended for use in low-income countries by the World Health Organisation. VA documents contain both coded data and open narrative. The task is formulated as a Text Classification problem and explores various combinations of linguistic and statistical approaches to determine how these may improve on the standard bag-of-words approach using a dataset of over 6400 VA documents that were manually annotated with cause of death. We demonstrate that a significant improvement of prediction accuracy can be obtained through a novel combination of statistical and linguistic features derived from the VA text. The paper explores the methods by which ATC may leads to improved accuracy in Cause of Death prediction.

Keywords: Verbal Autopsy, Cause of Death Prediction, Features, Text Classification.

1 Introduction

Not all deaths that occur annually are medically certified with Cause of Death (CoD). It is estimated that about 67 percent of the 57 million deaths that occur annually are not medically certified due to weak or negligible death registration systems, predominantly in low income countries[1].Information about CoD is a means to revealing preventable illness; developing health interventions; and research for treatment of diseases [2]. In low income countries there is pressure to find cost effective but still accurate CoD information and the Verbal Autopsy technique is frequently employed to do this [1].

The Verbal Autopsy (VA) technique is now well established in a large number of low income countries and generally follows the same pattern. It involves interviewing individuals (such as relatives or caregivers) who were close to the deceased, and if possible, those who cared for the individual around the time of death, in order to document events that may have led to the individual's death. The interviews are captured on a standard questionnaire or document that is then sent for analysis by physicians

I. Gurevych, C. Biemann, and T. Zesch (Eds.): GSCL 2013, LNAI 8105, pp. 47–60, 2013.
© Springer-Verlag Berlin Heidelberg 2013

who agree on a Cause of Death (CoD) classification based on the World Health Organisation (WHO) International Classification of Diseases (ICD) coding standards. It is worth noting that the VA interview is carried out in local languages of the countries in which they are employed, translated into English and transcribed onto the VA document for physicians to review.

Automatic prediction of CoD from the VA documents presents a number of benefits over the traditional manual physician based approach which is characterised by several limitations: high cost; intra-physician reliability; repeatability; inefficiencies and time consuming which automatic approaches may help overcome[3]. The potential benefits to be derived from the automated approaches to VA analysis and classification of CoD are attracting research interest [3, 4]. The VA document captures information of responses to both closed questions and open questions that record a narrative history. However, the automatic approaches published so far have only made use of the closed question responses [5, 6], while physicians have access to and make use of both the closed question responses and the open narrative in order to complete their diagnosis [7].

Our research is motivated by the hypothesis that computational algorithms that can take into account information obtained from the VA open narrative may lead to an improved prediction accuracy, which may in turn contribute to the United Nations' Millennium Development Goals. The research has been formulated as a Text Classification problem and classifies the VA according to CoD categories. We have employed Computational Linguistics (CL) and Machine Learning approaches to identify various features to be able to classify VA documents. We want to explore whether information derived from the open narrative leads to an improved CoD classification accuracy over either the closed question responses or a combination of both. Our literature review and experience in the field indicates that this is the first research that seeks to explore this approach.

Classification of biomedical documents is witnessing a high rate of growth in research in the applications of CL technology [8-11]. Cohen [11] for example employed chi-square as a statistical technique to extract features for a Support Vector Machine algorithm to classify genomes in biomedical text. Pakhomov et al [8] also employed various Text Classification based approaches to develop predictive models that identify patients with risk of heart failures from clinical notes obtained from Electronic Health Records.

The studies mentioned above have mainly explored the data originated from the formal environmental settings of the biomedical domain, where use of language is standardized with limited vocabulary. However, limited research has explored the informal settings where there are no specific rules but rather colloquial language is predominantly used. Nikfarjam and Gonzalez [12] and Leaman et al [13] are few researchers who have explored colloquial text within the biomedical domain. Nikfarjam and Gonzalez [12] employ CL approaches to automatically classify whether users experience adverse reactions of a given drug. Using data generated from Daily-Strength (www.dailystrength.org), they employed association rules to extract patterns of colloquial expressions that correlate with adverse reactions. Their work is largely motivated by the works in the area of automatic analysis and classification of sentiment and opinions, which are mostly expressed in colloquial text [14-17]. Pang et al [17] for example employed CL approaches to determine whether a sentiment expressed about a movie is positive, negative or neutral. Using various lexical and statistical features derived from a sample of movie review text, they demonstrate the possibility of using this approach with a comparable results obtained by humans.

Despite the emerging interest in research and numerous studies focused on automatic classification of colloquial text in general and specifically text from the biomedical, domain this has not been extended to VA text, which is argued by Danso et al [18] that the text should be considered a rather unusual subtype of biomedical genre. The next section gives a brief description of the VA dataset which we have used for our research and summarise the argument that VA it is distinct as a subtype of biomedical text.

2   Dataset

Danso et al[18] provide a detailed description of the dataset. In brief, unlike a standard biomedical text generated from a discourse either between a non-health professional and a health professional, or between health professionals, the VA text is a transcription of a discourse between two non-health professionals, written for a health professional (usually a medical doctor) to review. The dataset contains a total of over 11,700 VA documents of stillbirth, infants and women of reproductive age. It was collected from Kintampo in Ghana as part of a multi-year, single centre study between the year 2001 and 2010, and funded by the United Kingdom's Department for International Development. The VA text in this instance are electronic version of the interviews that were conducted in a local language called Twi, translated into English and transcribed onto the VA document by the non-medically trained interviewer. The dataset also contains the corresponding closed ended responses to each of the open narratives.

Figure 1and 2 below are shown to demonstrate the difference that exist between the closed response and open narrative as found in a typical VA document.

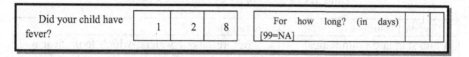

Fig. 1. Question and response options provided to respondent during interview

Can you tell me something about your pregnancy?
Movement of the baby in the womb started around the 6th month continuously till 9th month following the delivery. Although I did not encounter too many pregnancy complications, malaria persistently attacked me on the 7th month until I delivered. I suffered severely from anaemia which was diagnosed by a health worker when i visited hospital on the 8th month. Finally, I was not able to feed by self well when about a month to delivery due to lost of appetite. Sometime instead of feeding thrice a day, once daily becomes a problem for me.
Can you tell me something about your labour?
the labour started around 1pm in the night following the flow of water approximately 4hours. All of a sudden I felt the baby coming therefore I decided to try my best as much as possible to deliver at home. To my surprise the baby came with her both legs which really made it difficult to deliver myself. Therefore the TBA in the village was called to assist yet it proved futile. thus my husband had to go and arrange for vehicle to take me to the nearest hospital facility remarked by the TBA. before the vehicle arrived i had finally delivered.
Can you tell me something about the baby?
the baby landed without breathing or crying, therefore I enquired from the TBA to know what has happened to my baby but the woman assured me that the child is weak so I should lie down for a while and feel comfortable for everything will be alright. after she had finished with me she confirmed the baby landed dead.
Can you tell me what happened after delivery?
the baby neither cried or nor breath after delivery
Any signs and symptoms before the death of the child?
since the baby was very weak, he was put in an incubator but died after three hours of birth.

Fig. 2. A sample of open narrative question and responses from Infant Verbal Autopsy questionnaire

The above figures represent how both the open narrative and coded VA information are recorded. The open narrative is a verbatim account of the interview translated and transcribed by the interviewer and subsequently digitised. Figure 1 indicates the questions asked by the interviewer and the various options provided for the closed responses have the following meanings: 1="Yes", 2 = "No", 8 = "not known" and 99 = "Not application". The issues that characterised the text of the discourse of VA shown in figure 1 above are as catalogued in Table 1 below.

Table 1. Categories of issues with the VA dataset

Type of issue	example
A non-standard grammatical and spelling errors	"Before labour waters, which look clear and without bad scent" "… she fell sick, which lauted for three days.."
Colloquial forms in expressing concepts	Baby came out ⎫ Baby landed ⎬ Delivery Gave birth ⎭
Use of local terms to describe medical conditions	Asram, Anidane

Table 1. (*continued*)

A non-standard and fuzzy expressions of medical concepts	"I visited xxx hospital on Tuesday and was given one bottle of water. .."
Abbreviations and acronyms	TBA = Traditional Birth Attendant ANC = Antenatal care . CS = caesarean session
Inappropriate use of punctuation marks	"Any time, she breaths, you see a hole"

For this paper we report on the use of a subset of 6407 infant (stillbirths included) VA documents, which is approximately 1.6 million words in total, taken from the full dataset and used as the basis for the experiments being reported here. Each document in this subset has already been annotated by a minimum of two physicians and the final agreed CoD classification assigned. Where there is a disagreement between the two physicians, a third physician is employed to decide on the final CoD. There are two separate features of the CoD to be categorised: Time-of-Death has five categories and Type-of-Death consists of 16 categories as detailed in Tables 2 and 3 below.

Table 2. Breakdown of Time-of-Death categories

Time of Death	% distribution
Neonatal	31.3
Antepartum_stillbirth	21.5
PostNeonatal	19.1
Intrapartum_stillbirth	15.6
Non_stillbirth_unknown_cause	12.5

Table 3. Breakdown of Type-of-Death categories

Type of Death	% distribution
Stillbirth- unexplained	22.1
Cause unknown	12.5
Birth asphyxia	10.9
Neonatal Infection	8.7
Stillbirth-obstetric complications	7.5
PostNeonatal - Other Infections	6.9
Neonatal - other causes	5.9
Prematurity	5.8
Pneumonia	5.6
Malaria	4.3
Stillbirth- maternal disease	3.2
Diarrhoea	2.4
stillbirth- maternal haemorrhage	1.9
Stillbirth - other causes	1.5
stillbirth-congenital abnormalities	0.5
Measles	0.1

The issues outlined above characterise a dataset with high level of sparseness and lexical diversity [18]. To further demonstrate the relative noisy nature of the VA text, Danso et al [18] selected a sample of the VA text, which was used to evaluate the accuracy of a PoS tagger that was trained using the Brown corpus [19]. The evaluation of the performance of the PoS tagger carried out by a linguistic expert suggested an accuracy of 88 percent, which is a clear departure from the expected accuracy of about 96 - 97 percent from a normal English text [18].   Additionally, there are also issues about imbalance across CoD categories as shown in Tables 2 and 3 above. These issues present various levels of challenges for Computational Linguistics and Machine Learning based approaches to Text Classification and employing standard techniques in the area of biomedical Text Classification may not be produce desirable results.

3   Methodology

A brief description the methods employed in predicting CoD from VA is given here.

3.1   Evaluation

We employ the standard Precision, Recall and F1- measure metrics[20] to evaluate the performance of the classification methods against the physician CoD classification as a gold standard. Macro-averages as opposed to Micro-averages are used since Macro-averages tend to be suitable for highly skewed multi-class dataset which allows equal weights to be computed for each CoD category [21]. All averages are obtained based on 10 fold cross-validation [22].

3.2   Pre-processing

The texts were converted to lower case and tokenized by whitespaces. All punctuations were removed. Our initial exploration of the feature space pointed to the fact that removing the standard English stop-words affects tend to adversely affect the performance accuracy which falls below baseline of bag of words. All words were used in their natural form as they appeared and no further processing such as spell checking and corrections were carried out. With regards to the responses to closed questions part of the dataset, all information pertaining to symptoms, history of cares sought and treatment were extracted and separately stored. These were further discretised to ensure each category of response was appropriately used and not treated as a numeric value for the feature which has implications for the machine learning algorithm. For example question *"did your child have fever"* as indicated in figure 1 above, has three options 1="Yes", 2 = "No" and 8= "don't now" of responses and the numeric values are captured.

3.3   Classification Algorithm

Danso et al [23] previously showed in an experiment aimed at exploring the VA space in order to identify the suitable algorithm for this task. The results suggest Support Vector Machine (SVM) as the most suitable Machine Learning algorithm. The Sequential

Minimal Optimisation (SMO) algorithm, which is an implementation of SVM in WEKA Machine Learning software[24], was therefore used in this experiment.

3.4 Features for Classification

Danso et al [23] explored a baseline bag-of-words and how feature values must be represented for a classification algorithm. The results suggest a Normalised Term Frequency value as the best feature value representation. Normalized Term Frequency value was computed as the frequency of a given term, divided by the total number of terms in a given document. The experiments in this paper therefore employed the same scheme of representation. The various features that were explored in this experiment are as outlined as detailed in the table 4 below.

Table 4. List of features explored in this experiment

label	Features
A	Unigram (bag-of-words)
B	Unigrams + PoS Trigrams
C	Unigrams + PoS Trigrams + Relative Word positions
D	Unigrams + PoS Trigrams + Relative Word positions + Noun Phrases
E	Unigrams + PoS Trigrams + Relative Word positions + Noun Phrases + collocation bigrams (top 1 collocates)
F	Unigrams + PoS Trigrams + Relative Word positions + Noun Phrases + collocation bigrams (top 1 collocates + collocation bigrams (top 2 collocates)
G	Unigrams + PoS Trigrams + Relative Word positions + Noun Phrases + collocation bigrams (top 1 collocates + collocation bigrams (top 2 collocates)+ collocation bigrams (top 3 collocates)
H	Closed response
J	Closed response + Unigrams + PoS Trigrams + Relative Word positions + Noun Phrases + collocation bigrams (top 1 collocates + collocation bigrams (top 2 collocates)+ collocation bigrams (top 3 collocates)

Our motivations for employing the above features in our experiments are explained in turn below.

3.4.1 Linguistics Features

Building on the results obtained from the lexical features (bag-of-words) the following linguistic features have also been explored.

Part of Speech (PoS) information obtained from a part of speech tagger have been employed as features in several Text Classification works[. This approach is considered as a crude form of determining the correct sense of a given word in a text [25]. The PoS information were obtained using the Natural Language Processing Tool Kit's PoS tagger [26] trained using the Brown corpus [19] to tag the Verbal Autopsy dataset.

Part of Speech Trigrams: PoS tags have have been shown to be useful feature in numerous Text Classification problems. Gamon [14]for example, demonstrated the use of PoS trigrams in sentiment classification. We however explored various PoS tags in our experiment which include unigram, bigrams and trigrams of which PoS trigrams was found to be the best and results presented here.

Relative Word Position: our motivation to experiment this feature is based on one of the criticism of the famous bag-of-words approach to Text Classification that bag-of-words approach tends to ignore the order and syntactic relations between the words that occur in a sentence [27]. The adaption of the relative positions of words in text as possible features for Text Classification has been explored by various researchers [28, 29. Matsumoto et al [29] for example, demonstrate the usefulness of this feature by extracting word sub-sequences and dependency sub-trees from sentences to classify movie reviews. We however adapt a simplified approach by exploring the relative positions of the words as they appear in the text to capture the sequential order of events within the context of the VA.

Our approach treats the entire content of a VA document as a single string of words with an imaginary grid, where each cell represents a word which is a member of the string. Each cell is serially allocated a unique number and that represents the position of the word with respect to the entire string. The position number of the word captured is divided by the number of the string (number of cells) to obtain the relative word position with respect to the entire words in the VA document. The hypothesis here is that there may be a logical order of event in the history that led to the death of an individual, which may be a major factor in case profiling in an investigation process, and this feature may help in capturing this order. This is illustrated by an example, which is taken from a VA document.

"In the second month of the pregnancy,...labour started which I was at home in the morning..."

. If the order of these words is ignored one possible reading could be

"In the second month labour started...."

and this presents different scenarios and may mean a different outcome from medical perspective. The proposed relative word position features is to avoid this situation and preserve the in which the words appear in the VA document.

Noun Phrases: having obtained PoS information for every word in the text, chunking techniques as implemented by the regular expression below was used to extract noun phrases:

```
r'''NounPhrase:{<DT>?<JJ>*<NN>*}
```

The decision to explore these features was inspired from the fact that domain concepts are mostly expressed using multiword structures [30]. For example "*a normal labour*" was used to describe a type of labour a mother experienced during pregnancy, which is domain specific information. A generalised approach to capturing these types of mentions in the text is through extraction of noun phrases, which are derived from the PoS tagset and were represented as single terms.

3.4.2 Statistically Derived Collocation as Features

Statistically derived features are considered to be some form of phenomenon that tend to occur in the use of a language but that are not predictable. As observed by [31] that "*each word has a particular and roughly stable likelihood of occurring as argument, or operator, with a given word, though there are many cases of uncertainty disagreement among speakers, and change through time*". Collocation is one technique that can be used to capture the phenomenon described above. Collocation can be employed to capture word-pairs and phrases that frequently occur in the use of a language with no regard to their semantic or syntactic rules of use; and are also known to be dialect or language specific[32]. Collocation has been employed in many applications by lexicographers to carry out word sense disambiguation and semantic analysis of text[33].This therefore suggests an imperative investigation into the potential use of collocation as a feature to identify patterns of co-occurrence of words that could as indicative of phrase or an expressions of CoD considering the peculiar nature of colloquial text contained in the VA corpus.

We have employed statistical methods based on log-likelihood estimation to determine the likelihood of co-occurrence words and phrases [34]. The log-likelihood estimation was based on the entire corpus and estimated the likelihood of two words co-occurring as defined by bigram log-likelihood statistics association measure[35].The limitation associated with this approach however is that it usually take into account the only two word-collocate (bigrams) that co- occur in the corpus [36]. To address this limitation, we explored the levels of associations observed from the corpus as ranked by the bigram log-likelihood statistics association measure algorithm. The topmost collocation bigrams were experimented in turns and their impacts on performance were obtained. We illustrate our idea with the following example.

$$during \mid 'labour'= 4150, 'pregnancy'= 2901 \; 'my' = 1785$$

In the example shown above the word *during*, which is mentioned in a given VA document retrieves the three words with the strongest association with their corresponding likelihood values as ranked by the algorithms. These words are retrieved and added as part of the feature set.

3.4.3 Combined Feature Set

In order to explored whether the information obtained from both closed response and open narratives parts could improve classification of CoD in VA, all features derived from both parts where combined.

4   Results

The results presented here are based on the features described above. Many combinations were explored but for brevity only the best performing combinations are presented here. To give a better perspective of the results given here, we give the result obtained by a simple majority baseline algorithm ZeroR in WEKA as captured in table 4 below. We differentiate between time-of death   and type-of-death feature labels by adding ($_1$) and ($_2$) to the respective labels.

Table 5. Baseline results from a simple majority

	Category	Precision	Recall	F-measure
O_1	Time-of-Death	0.098	0.313	0.149
O_2	Type-of-Death	0.049	0.221	0.08

Table 6. Results from various feature sets - Time of Death categories

	Precision	Recall	F-measure
A_1	0.414	0.434	0.416
B_1	0.473	0.428	0.339
C_1	0.56	0.59	0.517
D_1	0.56	0.59	0.517
E_1	0.613	0.599	0.559
F_1	0.637	0.618	0.559
G_1	0.643	0.629	0.582

Table 7. Results from various feature sets- Type-of Death categories

	Precision	Recall	F-measure
A_2	0.248	0.288	0.251
B_2	0.22	0.256	0.142
C_2	0.314	0.376	0.285
D_2	0.314	0.376	0.285
E_2	0.33	0.391	0.304
F_2	0.311	0.395	0.306
G_2	0.35	0.406	0.322

Table 8. Results from various closed and narrative combined - Time of Death categories

	Precision	Recall	F-measure
H_1	0.826	0.836	0.827
J_1	0.826	0.835	0.828

Table 9. Results from various closed and narrative combined - Type of Death categories

	Precision	Recall	F-measure
H_2	0.575	0.616	0.583
J_2	0.591	0.616	0.587

5   Discussion and Future Work

The meaning of a word is best known by the context in which it exists and this was evident in the results obtained from these experiments. Using various linguistic and statically derived features which have extra information about the individual words has shown that there is significant increase in performance accuracy over the single words (bag-of-words) in predicting of CoD. Notable among these are our novel collocation and relative word positions features introduced. There were also marginal gains in terms of precision obtained from the PoS trigrams feature set. One feature set, which was however not found useful, is the noun phrase as shown in experiment G_1 and G_2 in table 6 and 7 respectively. The result remained unchanged when noun phrases were introduced. This is surprising considering our hypothesis about the potential usefulness of noun phrases in capturing multiword concepts. The result is however consistent with the findings in the literature that noun phrases tend not be useful features in ATC29] and require further investigations.

The results obtained from the combined set of features from both closed and open parts in Table 8 and 9 tend to suggest that the closed response part achieves better performance accuracy than the narrative part. The marginal gains in accuracy from the combined set in experiments J_1 and J_2 also tend to suggest that there is a marginal benefit in predicting CoD using a combination of the narrative and the closed responses. A detailed examination of the output at the individual level of CoD however suggests some benefits in combining information from both closed and narrative text. An example is 'intra-partum stillbirth' category where F-measure values of 0.36, 0.49 and 0.87 where recorded for narrative, closed and combined respectively suggesting the close response missed some relevant information from available in the narrative.

The relatively low F-measure recorded particularly for the Type-of-Death categories in either or both narrative and closed response parts of the VA suggest:

(i) Data skew: few samples for many CoDs, making them hard to classify with either or both of closed and open parts of the data; and

(ii) Complexity of diagnosis: some of the examples suggest that CoD is determined by the physicians using a complex combination of information from the overall VA document, and this is hard to capture in simplistic models used in Machine Learning classifiers.

However, considering the noisy and rather an unusual type of text being dealt with, there is the possibility that the features employed so far may not be effective enough in discriminating between CoDs. There is therefore the need for further exploration within the feature space of the narratives in order to increase the performance accuracies obtained. This may include adaptation of the standard PoS taggers for this particular type of text. This was clearly demonstrated by the PoS tagging experimental results obtained by Danso et al [18] as indicated in section 2. It may however be argued that the choice of Brown corpus for training the PoS tagger was inappropriate considering the difference in text, which may have resulted in the poor performance of the linguistic features extracted from the output of the PoS tagger. It must be pointed out that the choice was purely based on convenience as there was no linguistically annotated corpus readily available for both the VA and the biomedical domain.

Future work could therefore explore the possibilities of training the PoS tagger with corpus that has linguistic annotations from either the VA or biomedical domain. Future work will also explore features that will be targeted at the minority categories to increase the potential of improving the performance accuracies for these categories.

We believe this work may help reduce the cost and increase the accuracy of predicting CoD from VAs and therefore addresses vital global health challenges that confront developing countries in particular and the WHO at large.

Acknowledgement. We thank Professor Betty Kirkwood of the London School of Hygiene and Tropical Medicine and the entire members of the trial management team of the ObaapaVita and Newhints projects that generated the dataset used in carrying out this study. We furthermore thank the Director and Staff of Kintampo Health Research Centre, especially the head of the computer centre Mr Seeba Amenga-Etego and staff of the computer centre, who worked around the clock to get this corpus digitised. Our appreciation also goes to Centre for International Health and Development, University College London Institute of Child Health for their nomination of an award of a scholarship and the Commonwealth Scholarship Commission for their funding support for this research.

References

1. World Health Organization: WHO Handbook for Reporting Results of Cancer Treatments (WHO Offset Publication No. 48) (2004)
2. Kahn, K., Tollman, S.M., Garenne, M., Gear, J.S.: Validation and application of verbal autopsies in a rural area of South Africa. Tropical Medicine & International Health 5(11), 824–831 (2000)
3. Byass, P., Kathleen, K., Edward, F., Mark, A.C., Stephen, M.T.: Moving from Data on Deaths to Public Health Policy in Agincourt, South Africa: Approaches to Analysing and Understanding Verbal Autopsy Findings. PLoS Medicine 7(8) (2010)
4. King, G., Lu, Y., Shibuya, K.: Designing verbal autopsy studies. Population Health Metrics 8(1), 19 (2010)
5. Byass, P., Edward, F., Dao Lan, H., Yamene, B., Tumani, C., Kathleen, K., Lulu, M.: Refining a probabilistic model for interpreting verbal autopsy data. Scandinavian Journal of Public Health 34(1), 26–31 (2006)
6. Murray, C.J.L., Alan, D.L., Dennis, F., Shannon, T.P., Gonghuan, Y.: Validation of the symptom pattern method for analyzing verbal autopsy data. PLOS Medicine 4, 1739–1753 (2007)
7. Soleman, N., Chandramohan, D., Shibuya, K.: WHO Technical Consultation on Verbal Autopsy Tools, Geneva (2005)
8. Pakhomov, S., Shah, N., Hanson, P., Balasubramaniam, S., Smith, S.: Automatic quality of life prediction using electronic medical records. American Medical Informatics Association (2008)
9. Pakhomov, S., Weston, S.A., Jacobsen, S.J., Chute, C.G., Meverden, R., Roger, V.L.: Electronic medical records for clinical research: application to the identification of heart failure. The American Journal of Managed Care 13(6 Part 1), 281 (2007)
10. Cohen, A.M., Hersh, W.R.: A survey of current work in biomedical text mining. Briefings in Bioinformatics 6(1), 57–71 (2005)

11. Cohen, A.M.: An effective general purpose approach for automated biomedical document classification. In: AMIA Annual Symposium Proceedings. American Medical Informatics Association (2006)
12. Nikfarjam, A., Gonzalez, G.H.: Pattern mining for extraction of mentions of adverse drug reactions from user comments. In: AMIA Annual Symposium Proceedings. American Medical Informatics Association (2011)
13. Leaman, R., Wojtulewicz, L., Sullivan, R., Skariah, A., Yang, J., Gonzalez, G.: Towards internet-age pharmacovigilance: extracting adverse drug reactions from user posts to health-related social networks. Association for Computational Linguistics (2010)
14. Gamon, M.: Sentiment classification on customer feedback data: noisy data, large feature vectors, and the role of linguistic analysis. In: Proceedings of the 20th International Conference on Computational Linguistics, p. 841. Association for Computational Linguistics, Geneva (2004)
15. Oberlander, J., Nowson, S.: Whose thumb is it anyway?: classifying author personality from weblog text. In: Proceedings of the COLING/ACL on Main Conference Poster Sessions. Association for Computational Linguistics (2006)
16. Turney, P.D.: Thumbs up or thumbs down?: semantic orientation applied to unsupervised classification of reviews. In: Proceedings of the 40th Annual Meeting on Association for Computational Linguistics, pp. 417–424. Association for Computational Linguistics, Philadelphia (2002)
17. Pang, B., Lee, L.: Seeing stars: Exploiting class relationships for sentiment categorization with respect to rating scales. In: Annual Meeting-Association For Computational Linguistics (2005)
18. Danso, S., Atwell, E.S., Johnson, O., ten Asbroek, A., Soromekun, S., Edmond, K., Hurt, C., Hurt, L., Zandoh, C., Tawiah, C., Fenty, J., Etego, S., Agyei, S., Kirkwood, B.: A semantically annotated Verbal Autopsy corpus for automatic analysis of cause of death. ICAME Journal of the International Computer Archive of Modern English 37 (in press, 2013)
19. Francis, W.N., Kucera, H.: Brown corpus manual. Letters to the Editor 5(2), 7 (1979)
20. Scott, S., Matwin, S.: Text classification using WordNet hypernyms. In: Use of WordNet in Natural Language Processing Systems: Proceedings of the Conference (1998)
21. Forman, G.: A pitfall and solution in multi-class feature selection for text classification. ACM (2004)
22. Kohavi, R.: A study of cross-validation and bootstrap for accuracy estimation and model selection. Lawrence Erlbaum Associates Ltd. (1995)
23. Danso, S., Atwell, E.S., Johnson, O.: A Comparative Study of Machine Learning Methods for Verbal Autopsy Text Classification. International Journal of Computer Science Issues 10 (in press)
24. Witten, I.H., Frank, E.: Data Mining: Practical machine learning tools and techniques. Morgan Kaufmann (2005)
25. Pang, B., Lee, L., Vaithyanathan, S.: Thumbs up?: sentiment classification using machine learning techniques. In: Proceedings of the ACL-02 Conference on Empirical Methods in Natural Language Processing, vol. 10, Association for Computational Linguistics (2002)
26. Loper, E., Bird, S.: NLTK: the Natural Language Toolkit. In: Proceedings of the ACL-02 Workshop on Effective Tools and Methodologies for Teaching Natural Language Processing and Computational Linguistics, vol. 1, pp. 63–70. Association for Computational Linguistics, Philadelphia (2002)

27. Wilks, Y., Stevenson, M.: Word sense disambiguation using optimised combinations of knowledge sources. In: Proceedings of the 17th International Conference on Computational Linguistics, vol. 2. Association for Computational Linguistics (1998)
28. Moschitti, A., Basili, R.: Complex linguistic features for text classification: A comprehensive study. In: McDonald, S., Tait, J.I. (eds.) ECIR 2004. LNCS, vol. 2997, pp. 181–196. Springer, Heidelberg (2004)
29. Matsumoto, S., Takamura, H., Okumura, M.: Sentiment classification using word subsequences and dependency sub-trees. In: Ho, T.-B., Cheung, D., Liu, H. (eds.) PAKDD 2005. LNCS (LNAI), vol. 3518, pp. 301–311. Springer, Heidelberg (2005)
30. Scott, S., Matwin, S.: Feature engineering for text classification. In: Machine Learning-International Workshop Conference (1999)
31. Harris, Z.S.: Methods in structural linguistics (1951)
32. McKeown, K.R., Radev, D.R.: Collocations. Handbook of Natural Language Processing. Marcel Dekker (2000)
33. Pearce, D., Qh, B.: Using conceptual similarity for collocation extraction. In: Proceedings of the Fourth Annual CLUK Colloquium (2001)
34. Dunning, T.: Accurate methods for the statistics of surprise and coincidence. Computational. Linguistics 19(1), 61–74 (1993)
35. Seretan, V., Nerima, L., Wehrli, E.: Extraction of multi-word collocations using syntactic bigram composition. In: Proceedings of the Fourth International Conference on Recent Advances in NLP, RANLP-2003 (2003)
36. Pearce, D.: A comparative evaluation of collocation extraction techniques. In: Proceedings of the 3rd International Conference on Language Resources and Evaluation, LREC 2002 (2002)

SdeWaC – A Corpus of Parsable Sentences from the Web

Gertrud Faaß[1] and Kerstin Eckart[2]

[1] Institut für Informationswissenschaft und Sprachtechnologie
Universität Hildesheim, Hildesheim, Germany
gertrud.faass@uni-hildesheim.de
[2] Institut für Maschinelle Sprachverarbeitung
Universität Stuttgart, Stuttgart, Germany
kerstin.eckart@ims.uni-stuttgart.de

Abstract. For a number of languages, web crawling allows researchers to collect huge text samples to build corpora. However, only part of the material found on the internet is useful for Natural Language Processing, as e.g. parsers typically cannot handle lists and tables, or very short or very long sentences. There are methods (cf. e.g. [3]) for cleaning the downloaded data before adding it to a corpus collection – but even when these are applied, not all remaining textual material might be suitable for certain research requirements. This paper describes methods utilized to prepare *deWaC*, a freely available German web corpus of the WaCky project, for automatic processing up to the parsing level. It then discusses ways in which this corpus, called SdeWaC, has been used since its release.

1 Introduction

During the past years, a number of corpora collected from the World Wide Web have been made available to a wider public for free. These "gigaword corpora" [3] are of great use in Natural Language Processing (NLP) as they allow for data-driven research also on phenomena of comparatively low frequency. One such corpus is the German web corpus *deWaC* of the WaCky project [1, 2][1].

In the preparation of deWaC, four important steps have been applied after crawling, cf. [2]: A **pre-filtering** step that decides on the basis of mime type and size in kilobytes which documents stay in the collection and that also removes all instances of perfect document duplicates. A **cleaning** step that removes HTML and javascript code as well as boilerplate material[2] from the remaining documents. A **filtering** step that, among others, takes a list of function words into account to identify connected text and language. And a **near-duplicate removal** step based on the number of shared n-grams in a pair of documents.[3]

---

[1] Corpus available at http://wacky.sslmit.unibo.it/doku.php?id=corpora
[2] Recurring natural language text material like headers, disclaimers, etc., cf. [2].
[3] Note that, while most of the steps select or discard documents as a whole, the second step removes parts of document content.

I. Gurevych, C. Biemann, and T. Zesch (Eds.): GSCL 2013, LNAI 8105, pp. 61–68, 2013.

When using deWaC to extract data for NLP projects, however, we found a significant number of sentences that were not parsable[4], a high number of sentence duplicates, and some noise regarding the included top-level domains.

Therefore, we revised the complete source corpus in several steps. In the preparation of these steps and during sorting, some of the methods described by [11] were applied. To make the corpus usable for parsing-based extraction of linguistic data, e.g. [5, 20], we intended to identify parsable sentences and to keep only those. Our methodology has not yet been fully evaluated; we conducted a small study based on 200 sentences for first insight, see Sect. 2.5. However, we received positive feedback from the use cases and studies described in Sect. 3.

2   Methods and Implementation

We revised the corpus in several precision-oriented steps, i.e. we deleted sentences that carried a risk of being not processable by a parser, even when they showed no errors (e.g. very long sentences). This led to a corpus which was considerably reduced in size. Fig. 1 shows the workflow and the corpus size after each step.[5]

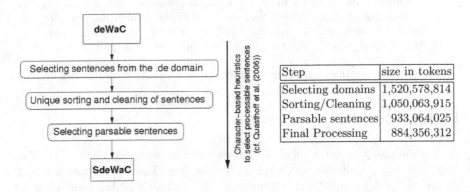

Step	size in tokens
Selecting domains	1,520,578,814
Sorting/Cleaning	1,050,063,915
Parsable sentences	933,064,025
Final Processing	884,356,312

Fig. 1. From deWaC to SdeWaC: workflow and token numbers

2.1   Selecting the Data by Domain

According to [1, 2], the deWaC crawl was limited to the .de and .at domains. However there is some noise, including top-level domains such as .it, .fr and .dk, or documents with generic top-level domains such as .com and .org; these data were excluded. For the preparation of SdeWaC, we only took into account data from .de domains, as for our research projects we need to differentiate between the different German national varieties. In selecting sentences depending on their origin, we relied on the work described in [19], where the data had been

---

[4] We regard a sentence not parsable, when our parser produces no or corrupted output.
[5] In the final processing step some pattern-based cleaning led to a further reduction.

sorted into .de vs. other domains[6]. This work also shortened the source information for each document to its domain name, i.e. up to the top-level domain. For ease of processing, the original data was re-formatted from the one-token-per-line format into a one-sentence-per-line format. Thereby, existing annotations were deleted. See Sect. 2.4 for the re-annotation step.

2.2 Sorting and Cleaning

Since we only require sentences which are parsable, and since we are interested in the frequency of occurrence of single words and word combinations, we sorted all sentences uniquely. However, since not only chunks, but also complete sentences might be formulated twice or more times by different authors or sources, duplicate sentences remained in the corpus if they were crawled from sources differing in their domain name. This processing step reduced the size of the corpus significantly, as several thousand sentences had had two or three duplicates. Obviously, the sorting and the removal of sentences led to a loss of contextually relevant data and makes the reconstruction of the original documents impossible.[7] The advantages of this procedure outweigh in our view its problems: we have a solid method for duplicate removal that allows us to run statistics, e.g. for word combinations, on the data. Furthermore, as downloading the original deWaC allows an interested user to retrieve sentences of interest with their deWaC context, we accepted the lack of context.

For cleaning, we followed most parts of [11]'s approach by e.g. examining the number of non-alphabetic characters in relation to the total number of characters in a sentence, in order to identify non-sentential character sequences.

2.3 Finding and Deleting Sentences Not Suitable for Parsing

The output of many parsers is a fully disambiguated syntactic tree. However, this disambiguation often involves forced guessing at a specific processing step and may lead to a corrupted syntactic analysis for an otherwise syntactically well-formed sentence. To avoid this, we applied the FSPar [12] parser to identify the expected parse quality, as it does not force disambiguation but encodes underspecified results in its output graphs. This way no sentences were lost due to forced disambiguation decisions.

FSPar is a rule-based dependency parser that uses a huge lexical resource and includes its own preprocessing chain, i.e. a tokenizing step and tagging by the TreeTagger [14].[8] As we encountered some problems due to the character encodings found in deWaC, we applied a Latin-1 encoding for the parser input.[9]

---

[6] For a number of sentences contained in the de-part of [19], the domain information was lost, however, we kept these sentences as well.

[7] Note, that the cleaning step in the preparation of deWaC, described in [2], also removes parts of document content and makes use of heuristics for boilerplate stripping which may have an effect on the original context.

[8] On a 2.8 GHz core a sentence is processed by the FSPar pipline in about 2ms.

[9] Non-Latin-1-characters were encoded as numerical HTML encoding of their unicode character, to allow for a re-formatting into UTF-8.

FSPar's output, cf. Fig. 2[10] and Table 1, is based on the CoNLL 2006 format [6]. To find sentences not suitable for further processing, we make use of column 6. It contains "-1" to signal the main verb of the sentence (root of the dependency graph) and some punctuation signs. It also uses this notation to signal those words which it cannot integrate into a dependency structure. We determine an *error rate* for each sentence, by calculating the relative frequency of all its "-1" occurences, taking into account that the main verb and the punctuation symbol at the end of a sentence are correctly assigned "-1". This error rate helps us to decide how well the sentence could be parsed by FSPar. The parsability assessment could obviously have profited from the use of different parsers, but as no other parser producing such output was available, we relied on FSPar's results only. The sentences with the worst results (error rate above 0.7) were deleted. For the evaluation, see Sect. 2.5.

```
<article cluster=10475 address=0 date=2007>
<s>
0   Warum         PWAV   warum          |                              9/1  ADJ
1   sind          VAFIN  seinA          1:Pl:Pres:Ind|3:Pl:Pres:Ind -1  TOP
2   in            APPR   in             Dat                            9/1  ADJ|PP/in4|ADJ
3   der           ART    d              |                              4    SPEC
4   M"uslipackung NN     M"usli#@packung Dat:F:Sg                      2    PCMP
5   die           ART    d              |                              7    SPEC
6   dicken        ADJA   dick           |                              7    ADJ
7   N"usse        NN     Nuss           Nom:F:Pl                       9/1  NP:1
8   immer         ADV    immer          |                              9/1  ADJ
9   oben          ADV    oben           |                              9/1  RK
10  nach          APPR   nach           Dat                            9/1  ADJ|ADJ|PP/nach:4
11  dem           ART    d              |                              12   SPEC
12  Sch"utteln    NN     Sch"utteln     Dat:M:Sg|Dat:N:Sg              10   PCMP
13  ?             $.     ?              |                              -1   TOP
</s>
</article>
```

Fig. 2. Output of FSPar, tabs and special characters slightly reformatted for readability Example: Why are the thick nuts always on top after shaking the cereal box?

We are aware of the fact that one parse, generated by one parser, is only a weak decision criterion when it is necessary to decide whether a sentence is parsable in general. However, our goal was precision-oriented, and experience with the data showed that most of the sentences retained with this method were of a good quality and parsable by other parsers, as well. So far, SdeWaC has also been successfully parsed by a data-driven state-of-the-art dependency parser [4].

2.4 Enriching the Corpus with Linguistic Information

The resulting corpus was reformatted to UTF-8, processed with a tokenizer [15] and tagged/lemmatized with the TreeTagger [14]. Then some known errors of the tokenizer were corrected, e.g., wrong disambiguations of the period in numbers.

---
[10] Part-of-speech tagset: STTS, cf. [13]; roles: ADJ – adjunct, NP:1 – subject, PCMP – complement of a preposition or conjunction, PP/[in|nach]:4 – subcategorized prepositional phrase in dative case, RK – right sentence bracket, SPEC – specifier.

Table 1. An output format of FSPar (Col. 6 and 7 can appear repeatedly for a token)

Column	Content
1	token no./sentence border/ article border
2	token (full form)
3	part-of-speech tag
4	lemma
5	morphological information (underspecified)
6	head (underspecified)
7	role (underspecified)
8	pronoun resolution (underspecified)

Table 2. 200 sentence evaluation

manual error rate	bad	indifferent	good
(i) $\geq 0.7 (\pm split)$	5	5	5
(ii) $< 0.7 + split$	9	4	2
(iii) $< 0.7 - split$	5	23	142

2.5 Evaluating Our Method

A sample of 200 sentences of the result set after the sorting and cleaning step was randomly chosen for manual inspection. To gain first insight, we checked each of the sentences whether we would have accepted it as correct and compared it to the result of the automatic selection. Three categories ('good','bad','indifferent') were assigned by two annotators. Due to the 'indifferent' category and the fact that there are no explicit distinctions for sentence quality the annotators only reached moderate agreement [10] ($\kappa = 0.53$). Table 2 shows the results of the evaluation.[11] FSPar is able to split what was considered one input sentence into two or more sentences, thereby creating e.g., two syntactically sound parts from cases where a title was attached to the first sentence. Thus the error rate is assigned to each of the split parts and the evaluation categories regarding the input sentences in Table 2 are as follows: (i) sentences of which at least one part was rated ≥ 0.7 ('bad'), (ii) split sentences of which all parts were rated < 0.7 ('good') and (iii) non-split sentences, rated < 0.7 ('good'). As expected in the precision-oriented approach, some ($1/3$) of the sentences ruled out by the error rate were considered 'good' by the annotators. However 5 of the 19 sentences the annotators considered 'bad' were neither ruled out, nor split by the parser. In two cases a word started in upper or lower case where it should have been the other way around. However the other three sentences were not syntactically sound, as already the input sentence was corrupted due to a punctuation sign in mid-sentence, cf. Ex. (1).

(1)   Zudem   verdeutlicht der Tarifbegriff in § 63 Unterabs .
        Moreover clarifies   the rate notion in § 63 Subsect   .

---

[11] Cases in which the annotators did not agree were included as 'indifferent'.

While the robustness of the parser is rather positive in the first case, the decision regarding the other three sentences suffers from the fact that the parser is not able to signal missing information.

3   SdeWaC Use Cases

Since the release, several use cases and corpus linguistic studies have been conducted based on SdeWaC: [7] combine syntactic and semantic analyses for a semantic disambiguation of sortally ambigious -*ung*-nominalizations. In this project a significant number of parsable sentences containing -*ung*-nominalizations of verbs from a specific verb class were extracted from SdeWaC. [21] present a corpus study on passive forms of reflexive verbs in German. As this is a low frequency phenomenon which can only be automatically extracted by taking syntactic information into account, SdeWaC provided an adequate corpus resource for this study. [9] showed that applying a data-driven state-of-the-art dependency parser [4] to web data significantly decreases its scores. However they combined two parsed versions of SdeWaC to test linguistic hypotheses about the syntax-semantics interface of German *nach*-particle verbs on "real world" data. [18] utilized SdeWaC to extract empirical features for classification experiments. They combined theoretical and empirical perspectives towards an automatic disambiguation of German *an*-particle verbs.

SdeWaC was also used for testing the coverage of NLP tools, such as the Morphological Analyzer SMOR [16], evaluated by [8]. For this purpose, parsable sentences are not necessary, as only single words are extracted from the corpus. However, cleaning the corpus beforehand has the positive side effect that fewer non-words and fewer words with erroneous orthography appear in candidate lists used to test the coverage of the morphology system.

And lastly after testing compositionality of nouns based on their cooccurence in several corpora, [17] concludes that when acquiring such lexico-semantic information, a well-prepared web corpus like SdeWaC is an appropriate resource.

4   Results

The resulting corpus is kindly made available by the WaCky-project [2][12]. It comes in two formats: (i) one-sentence-per-line, and (ii) tokenized, tagged and lemmatized. The metadata *year*[13], *source*, and *error rate* are provided for each sentence ("0" denoting non-available information). Format (ii) can also be used as a basis to encode the corpus for OCWB (Open Corpus Work Bench)[14] and the CQP query language. The domain names of the source urls are not stored directly in the dataset, but are encoded numerically, cf. [19]. A list containing all known domain names and their respective codes is provided with the corpus.

---

[12] http://wacky.sslmit.unibo.it

[13] The year information was extracted from those source urls, where parts of the string seemed to denote a year [19].

[14] http://cwb.sourceforge.net/

References

1. Baroni, M., Kilgarriff, A.: Large linguistically-processed web corpora for multiple languages. In: Conference Companion of EACL 2006, 11th Conference of the European Chapter of the Association for Computational Linguistics, pp. 87–90 (2006)
2. Baroni, M., Bernardini, S., Ferraresi, A., Zanchetta, E.: The WaCky Wide Web: A Collection of Very Large Linguistically Processed Web-Crawled Corpora. Language Resources and Evaluation 43(3), 209–226 (2009)
3. Bauer, D., Degen, J., Deng, X., Herger, P., Gasthaus, J., Giesbrecht, E., Jansen, L., Kalina, C., Krüger, T., Märtin, R., Schmidt, M., Scholler, S., Steger, J., Stemle, E., Evert, S.: Fiasco: Filtering the internet by automatic subtree classification. In: Fairon, C., Naets, H., Kilgarriff, A., de Schrvyer, G.-M. (eds.) Building and Exploring Web Corpora: Proceedings of the 3rd Web as Corpus Workshop (WAC3), Incorporating CLEANEVAL, Louvain-la-Neuve, Belgium, pp. 111–121 (2007)
4. Bohnet, B.: Top accuracy and fast dependency parsing is not a contradiction. In: Proceedings of the 23rd International Conference on Computational Linguistics (Coling 2010), Coling 2010 Organizing Committee, Beijing, China, pp. 89–97 (2010)
5. Briscoe, T., Carrol, J.: Automatic extraction of subcategorization from corpora. In: Proceedings of the Fifth Conference on Applied Natural Language Processing, Washington DC, USA, pp. 356–363 (1997)
6. Buchholz, S., Marsi, E.: CoNLL-X Shared Task on Multilingual Dependency Parsing. In: Proceedings of the Tenth Conference on Computational Natural Language Learning (CoNLL-X), pp. 149–164. Association for Computational Linguistics, New York City (2006)
7. Eberle, K., Faaß, G., Heid, U.: Proposition oder Temporalangabe? Disambiguierung von -ung-Nominalisierungen von verba dicendi in nach-PPs. In: Chiarcos, C., Eckart de Castilho, R., Stede, M. (eds.) Proceedings of the Biennial GSCL Conference 2009, Von der Form zur Bedeutung: Texte automatisch verarbeiten / From Form to Meaning: Processing Texts Automatically, Potsdam, pp. 81–91. Narr, Tübingen (2009)
8. Faaß, G., Heid, U., Schmid, H.: Design and application of a Gold Standard for morphological analysis: SMOR in validation. In: Proceedings of the Seventh LREC Conference, European Language Resources Association (ELRA), Valetta, Malta, pp. 803–810 (2010)
9. Haselbach, B., Eckart, K., Seeker, W., Eberle, K., Heid, U.: Approximating Theoretical Linguistics Classification in Real Data: the Case of German "nach" Particle Verbs. In: Proceedings of COLING 2012, pp. 1113-1128. The COLING 2012 Organizing Committee. Mumbai, India (2012)
10. Landis, J.R., Koch, G.G.: The measurement of observer agreement for categorical data. Biometrics 33(1), 159–174 (1977)
11. Quasthoff, U., Richter, M., Biemann, C.: Corpus portal for search in monolingual corpora. In: Proceedings of the LREC 2006, Genoa, Italy, pp. 1799–1802 (2006)
12. Schiehlen, M.: A Cascaded Finite-State Parser for German. In: Proceedings of the Research Note Sessions of the 10th Conference of the European Chapter of the Association for Computational Linguistics (EACL 2003), Budapest, pp. 163–166 (2003)
13. Schiller, A., Teufel, S., Thielen, C.: Guidelines für das Tagging deutscher Textcorpora mit STTS. Universität Stuttgart and Universität Tübingen (1995)
14. Schmid, H.: Probabilistic Part-of-Speech Tagging Using Decision Trees. In: International Conference on New Methods in Language Processing, Manchester, UK, pp. 44–49 (1994)

15. Schmid, H.: Unsupervised Learning of Period Disambiguation for Tokenisation. Internal Report, IMS. University of Stuttgart (2000)
16. Schmid, H., Fitschen, A., Heid, U.: SMOR: A German computational morphology covering derivation, composition, and inflection. In: Proceedings of LREC 2004, Lisboa, Portugal (2004)
17. Schulte im Walde, S.: Webkorpora für die automatische Akquisition lexikalisch-semantischen Wissens. In: Workshop Webkorpora in Computerlinguistik und Sprachforschung. Institut für Deutsche Sprache, Mannheim (2012)
18. Springorum, S., Schulte im Walde, S., Roßdeutscher, A.: Automatic Classification of German an Particle Verbs. In: Proceedings of the 8th International Conference on Language Resources and Evaluation. Istanbul, Turkey (2012)
19. Stus, O.: Web-Korpus, Korpusaufbereitung der deutschen Web-Korpora. Internal Report, IMS, Universität Stuttgart (2008)
20. Weller, M., Heid, U.: Extraction of german multiword expressions from parsed corpora using context features. In: Calzolari, N., Choukri, K., Maegaard, B., Mariani, J., Odijk, J., Piperidis, S., Rosner, M., Tapias, D. (eds.) Proceedings of the Seventh Conference on International Language Resources and Evaluation (LREC 2010), pp. 3195–3201. European Language Resources Association (ELRA), Valetta (2008)
21. Zarrieß, S., Schäfer, F.: Schulte im Walde, S.: Passives of reflexives: a corpus study. Linguistic Evidence - Berlin Special. Berlin, Germany (2013)

Probabilistic Explicit Topic Modeling Using Wikipedia

Joshua A. Hansen, Eric K. Ringger, and Kevin D. Seppi

Department of Computer Science,
Brigham Young University, Provo, Utah, USA 84602

Abstract. Despite popular use of Latent Dirichlet Allocation (LDA) for automatic discovery of latent topics in document corpora, such topics lack connections with relevant knowledge sources such as Wikipedia, and they can be difficult to interpret due to the lack of meaningful topic labels. Furthermore, the topic analysis suffers from a lack of identifiability between topics across independently analyzed corpora but also across distinct runs of the algorithm on the same corpus. This paper introduces two methods for probabilistic *explicit* topic modeling that address these issues: Latent Dirichlet Allocation with Static Topic-Word Distributions (LDA-STWD), and Explicit Dirichlet Allocation (EDA). Both of these methods estimate topic-word distributions *a priori* from Wikipedia articles, with each article corresponding to one topic and the article title serving as a topic label. LDA-STWD and EDA overcome the nonidentifiability, isolation, and unintepretability of LDA output. We assess their effectiveness by means of crowd-sourced user studies on two tasks: topic label generation and document label generation. We find that LDA-STWD improves substantially upon the performance of the state-of-the-art on the document labeling task, and that both methods otherwise perform on par with a state-of-the-art *post hoc* method.

1 Introduction

The management and utilization of massive datasets is one of the great challenges of our time. The emergence of the Internet as a mass technocultural phenomenon has been accompanied by an explosion in textual data in particular. These new data sources present ripe fruit for all manner of analysis, with insights in linguistics, anthropology, sociology, literature, organizational behavior, economics, and many other areas of human endeavor merely waiting to be discovered. However, our own human limitations stand as the chief obstacle to advances at this scale. Even the most voracious human mind could never hope to take in more than the minutest fraction of this endless informatic ocean. Fortunately, the same digital systems that initially fostered the seemingly endless flood of data also prove useful in taming the deluge. One computational tool increasingly used to understand large unstructured datasets is probabilistic topic modeling, which "enables us to organize and summarize electronic archives at a scale that would be impossible" by human effort alone [1].

I. Gurevych, C. Biemann, and T. Zesch (Eds.): GSCL 2013, LNAI 8105, pp. 69–82, 2013.
© Springer-Verlag Berlin Heidelberg 2013

The well-known Latent Dirichlet Allocation (LDA) is a probabilistic, generative model of document collections that models documents as mixtures over latent topics [2], where topics are categorical distributions over some vocabulary. These topics can be discovered without supervision using any of a number of inference methods and are useful for exploring document corpora in terms of the themes or topics that they contain [1],[7].

What follows is an example of a subset of the topics discovered by LDA in a corpus of State of the Union messages, annual speeches given by presidents of the USA over the course of the country's history.[1]

Topic 0	Topic 1	Topic 2	Topic 3	Topic 4
defense	act	post	government	world
military	congress	service	great	peace
forces	law	mail	country	nations
strength	bill	department	means	nation
security	session	postal	experience	war

This example illustrates the ability of a topic model to discover unifying themes across presidents, across speeches, and across time. In this case, those patterns largely correspond to what is known about the historical circumstances under which the documents in the corpus were created. Furthermore, the example illustrates the ability of topic models to summarize document collections in terms of the topics discussed therein. This compact analysis provides a level of semantic abstraction in many ways more instantly usable by human users and analysts than the documents themselves.

Topics discovered by latent topic models such as LDA usually do not originate in the corpus in question but were already in use in other contexts by users of the language. Thus latent topic models often simply *re*discover topics already discussed elsewhere, while being completely ignorant of those pre-existing topics. When latent topic models encounter the same topic in different corpora, the implied linkage between the different instantiations of the topic can only be discovered by time-consuming *post hoc* comparisons. We term this the *inter-run identifiability problem*. This problem is related to the phenomenon of label switching in stochastic inference for mixture models, in which the composition of a topic is not guaranteed to persist across sampler iterations. This property also extends across iterations produced in distinct runs of such inference.

Topics generated by LDA and other latent topic models are also unsatisfactory in their lack of labels, thus requiring *post hoc* labeling for easier human interpretation. Fast, automatic methods, such as concatenating the top N words per topic, produce labels that over-represent high-weighted terms and are often hard to interpret. Manual labeling may require substantial human effort to understand the contexts of usage and introduces obvious subjectivity. Also, better automatic methods for labeling currently require separate, often complicated processing in addition to the original topic modeling.

---

[1] http://bit.ly/185kfmw, http://bit.ly/ZuJfO2, http://bit.ly/13nOb9R

This paper describes two approaches to overcoming LDA's deficiencies in inter-run identifiability and topic labeling. Both approaches involve adapting LDA for use with *explicit* pre-existing topics, rather than *latent* topics. The first approach, known as Latent Dirichlet Allocation with Static Topic-Word Distributions (LDA-STWD), simply substitutes precomputed topic-word counts into the standard LDA complete conditional distribution, allowing inference by Gibbs sampling to be used essentially unmodified. The second approach, known as Explicit Dirichlet Allocation (EDA), is a new model similar to LDA but is rederived from first principles after defining topic-word counts as explicit (and without a Dirichlet prior). In both of these approaches, the explicit topics are estimated in advance from Wikipedia articles, one topic per article. Defining topics in this way provides comparability of topic model output across corpora and provides human-defined labels for all topics in the form of the article title. In this paper we use this latter property to implement a document labeling algorithm that outperforms the current state of the art. We also demonstrate that probabilistic topic models incorporating pre-specified topics with their accompanying labels can assign labels to topics and documents that reflect the meaning of the topics and the content of the individual documents more effectively than a combination of the LDA topic model and a *post hoc* labeling method.

2   Related Work

In this section we situate the new LDA-STWD (Section 3.1) and EDA (Section 3.2) models within the space of topic modeling. Since these models will be evaluated (Section 4) using a document labeling task in which document labels are generated from the topic label of the document's most prominent topic, we also review the literature on automatic topic labeling.

2.1   Topical Representations of Documents

Topic models can be categorized according to whether the topic-document relationship is discovered using non-probabilistic techniques such as singular value decomposition (SVD) or through probabilistic inference methods, and according to whether topics are treated as latent and awaiting discovery or as pre-specified and explicit. A typology of prominent existing approaches—as well as of the approaches described in this paper—is given in Table 1.

Table 1. A typology of topical representations of documents

	Latent Topics	Explicit Topics
Non-Probabilistic	LSA[†]	ESA[‡]
Probabilistic	PLSA[§], LDA[♭]	LDA-STWD[♯], EDA[♭]

Key to Abbreviations: [†]Latent Semantic Analysis [‡]Explicit Semantic Analysis [§]Probabilistic Latent Semantic Analysis [♭]Latent Dirichlet Allocation [♯]Latent Dirichlet Allocation with Static Topic-Word Distributions (introduced in Section 3.1). [♭]Explicit Dirichlet Allocation (introduced in Section 3.2)

Latent Semantic Analysis [5] discovers topics or "concepts" by constructing a term-document matrix and reducing its dimensionality using singular value decomposition. This method results in term and document representations in a latent concept space. In Probabilistic Latent Semantic Analysis [9], topics are still latent, but rather than use the tools of linear algebra to discover them, PLSA recasts LSA as a statistical inference problem. PLSA is probabilistic but not fully generative; i.e., the model cannot be used to generate new documents. In the model's partial generative process, the latent document-topic and topic-word distributions are estimated using an expectation maximization algorithm. The authors observe that by explicitly modeling "contexts" (topics), the model can better represent polysemy beyond the ability of LSA to do so.

Latent Dirichlet Allocation [2] is likewise a probabilistic counterpart to LSA with latent topics, but unlike PLSA it is fully generative. The model specifies a Dirichlet prior distribution over the per-document mixing proportions $P(\theta \mid \alpha)$ and another over the topic-word distributions $P(\phi \mid \beta)$. In addition to producing a fully generative model that can generate new documents and compute probabilities of previously unseen documents, LDA also consistently outperforms PLSA in empirical evaluations carried out by its creators. As a result, LDA has become the basis for a myriad of derivative models.

Explicit Semantic Analysis [6] represents documents in Wikipedia-derived semantic space by ranking each Wikipedia concept (corresponding to an article) by how well its terms are represented within a given document. Selected results appear impressive, though evaluation is conducted only in terms of a word-relatedness task. The algorithm is not formulated probabilistically—indeed, the authors make no mention of probabilistic approaches whatsoever—but may lend itself to a probabilistic reinterpretation.

One analysis of LSA, LDA, and ESA finds significant benefit in ESA from the use of explicit topics [4]. Explicit topics are intuitively appealing because they provide a semantic summary of documents in terms of human-defined concepts rather than machine-discovered topics. Yet ESA assumes that a document consists of a single topic by seeking Wikipedia concepts that maximize the relatedness of the document as a whole, whereas documents are more effectively modeled as mixtures of topics, as in LDA, to be discovered jointly. This paper addresses this issue by placing explicit topic corpus modeling in a probabilistic framework.

Labeled LDA is a variation on LDA in which document topics are assigned *a priori* [15]. The authors apply it to a corpus of web pages with labels taken from the *del.icio.us* social bookmarking site. The essential nature of the model is that topic-word distributions remain latent, while document-topic distributions are observed—a complementary approach to ours, which is to treat topic-word distributions as observed while document-topic distributions remain latent.

2.2   Topic Labeling

Topic labeling is the task of automatically generating a label for a given topic, defined as a categorical distribution over the elements in some vocabulary. These labels are intended to aid human interpretation of topic model output, and a number of approaches to topic labeling have been described. Mei, Shen, and Zhai first defined the topic labeling problem as the production of "a sequence of words which is semantically meaningful and covers the latent meaning of [the distribution over words in a topic]" [14]. The authors also describe a typical approach to labeling: generate candidate labels, rank them according to some measure of relevance, and select a high-scoring candidate or candidates to serve as the label. Their candidates are generated from a corpus of relevant text, and are scored using various association measures computed on that corpus. Human annotators rated the quality of the selected high-scoring candidate labels, as well as the quality of baseline labels consisting of the concatenation of the top k words. Humans preferred labels from the new method over those of the baseline.

Magatti, Calegari, Ciucci et al. [13] vary this approach by taking candidate labels from the Open Directory Project hierarchy. Rather than choosing highly ranked candidates according to a single relevance score, they employ a complex scheme for combining the output of multiple relevance measures into a single answer. Unfortunately, they do not empirically validate the quality of the labels generated by their method.

Spitkovsky and Chang generate a mapping from phrases to Wikipedia URLs and back by observing the frequency of various anchor text phrases on hyperlinks pointing to Wikipedia pages [17]. The forward mapping could be applied to derive a distribution over Wikipedia article topics given the text of a document, but the authors do not develop such a method.

Lau, Newman, Karimi, et al. [11] follow much the same approach as Mei, Shen, and Zhai but define the task more narrowly as selecting a subset of the words contained in a topic, from which a label can be generated. They also introduce a supervised approach with accompanying performance gains. Lau, Grieser, Newman, et al. [10] extend this work by generating label candidates from relevant Wikipedia article titles as well as from top words in a topic. Association measures computed on the entire English Wikipedia are then used to rank the candidates for final label selection. By outperforming the Mei, Shen, and Zhai approach (and presumably that of Lau, Newman, Karimi, et al. on account of generating candidate labels that are a superset of those in that paper) this approach distinguishes itself as the current state of the art in automatic topic labeling and will serve as a point of comparison in our experiments.

In all of the above approaches, topic labels are generated *post hoc*. By contrast, our methods Latent Dirichlet Allocation with Static Topic-Word Distributions and Explicit Dirichlet Allocation are both constructed to provide integrated approaches in which the topic label is included in the output of the topic model itself.

3   Probabilistic Explicit Topic Modeling

In this section we describe two probabilistic explicit topic models. The first is Latent Dirichlet Allocation with Static Topic-Word Distributions (LDA-STWD), which is identical to LDA with the exception that topic-word distributions are pre-specified. The second is Explicit Dirichlet Allocation (EDA) which is a probabilistic explicit topic model derived from first principles.

3.1   Latent Dirichlet Allocation with Static Topic-Word Distributions (LDA-STWD)

We adapt LDA to use observed topic-word distributions. Document-topic distributions remain latent. As a first, albeit *ad hoc* attempt, we adapt the Gibbs sampler for LDA developed by Griffiths and Steyvers [8] (see eq. 5 of Griffiths and Steyvers) to model corpora in terms of *predefined* topic-word distributions. The change reformulates the LDA complete conditional distribution by directly replacing the topic-word counts (from the *target corpus*, the corpus to which the model is being applied) with those computed from a chosen *topic corpus* (such as Wikipedia). Unlike inference with latent topics, these new counts do not vary during the course of sampling. We call this first approach Latent Dirichlet Allocation with Static Topic-Word Distributions. The LDA Bayesian network and conditional probability distributions are otherwise unchanged; nevertheless, the d-separability implications of rendering ϕ observed are ignored. The complete conditionals for the new model are as follows:

$$P\left(z_{ij} = k \mid \overline{z}_{\neg ij}, \overline{\overline{w}}\right) \propto \frac{{}_\star\lambda_{ij}^k + \beta}{{}_\star\lambda_k^\star + J\beta} \cdot \frac{{}_i n_k^\star + \alpha}{{}_i n_\star^\star + K\alpha} \tag{1}$$

where symbols are defined as in Figure 1.

The corresponding complete conditional distribution for LDA employs ${}_\star n_k^{ij}$ and ${}_\star n_t^\star$ in place of ${}_\star\lambda_{ij}^k$ and ${}_\star\lambda_t^\star$, respectively.

The resultant model insists that the documents in the target corpus were generated from a set of pre-existing topics and topic word distributions. The sampler is thus expected to allocate the token-level topic assignments amongst those topics that best describe the corpus. In other words, the sampler simply chooses *which topics* to use, but not which words the topics will contain.

3.2   Explicit Dirichlet Allocation

Now we derive a purpose-built probabilistic explicit topic model Explicit Dirichlet Allocation, another adaptation of LDA for use with predefined, explicit topics. In EDA, documents are still modeled as a probabilistic admixture of topics, where a topic is a categorical distribution over words in a vocabulary. However, the topics (ϕs) in EDA are treated as observed or explicit, whereas in LDA they are considered unobserved or latent. Additionally, EDA's ϕ is no longer conditioned on a parameter β.

- $_\star\lambda_{ij}^k$ is the number of times words of the same type as token w_{ij} are assigned to topic k in the topic corpus (excluding token w_{ij} itself),
- $_\star\lambda_t^\star$ is the number of times topic t is assigned to any word in the topic corpus (excluding token w_{ij} itself),
- $\bar{\bar{z}}_{\neg mn}$ denotes the set of all token topic variables except z_{mn}.
- $_in_{k\neg mn}^\star$ denotes the number of tokens in document i assigned to topic k, with the exception of token w_{mn} (if it occurs in the document)
- $_\star n_k^{ij}$ is the number of times words of the same type as token w_{ij} are assigned to topic k excluding the current assignment of z_{ij} from the counts,
- $_in_k^\star$ is the number of tokens in document i assigned to topic k
- $_\star n_t^\star$ is the number of times topic t is assigned to any word excluding the current assignment of z_{ij} from the counts,
- $_in_\star^\star$ is the number of tokens in document i
- J is the total number of tokens [in the topic corpus] ·
- M is the number of documents in the target corpus
- K is the number of topics [in the topic corpus]

Fig. 1. Definitions of symbols

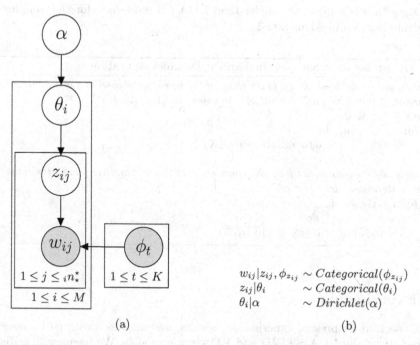

(a)

$$w_{ij}|z_{ij}, \phi_{z_{ij}} \sim Categorical(\phi_{z_{ij}})$$
$$z_{ij}|\theta_i \quad\quad \sim Categorical(\theta_i)$$
$$\theta_i|\alpha \quad\quad\quad \sim Dirichlet(\alpha)$$

(b)

Fig. 2. Graphical model (a) and conditional probability distributions (b) for Explicit Dirichlet Allocation

Definitions of the symbols are shown in Figure 1. The distribution of interest is the collapsed joint distribution over tokens and topics (the derivation has been omitted due to length limitations):

$$P\left(\overline{\overline{w}}, \overline{\overline{z}} \mid \phi, \alpha\right) = \prod_{i=1}^{M} P\left(\boldsymbol{w}_i, \boldsymbol{z}_i \mid \phi, \alpha\right) = \prod_{i=1}^{M} \int_{\theta_i} P\left(\boldsymbol{w}_i, \boldsymbol{z}_i, \theta_i \mid \phi, \alpha\right) d\theta_i \qquad (2)$$

$$= \frac{\Gamma\left(\sum_{k=1}^{K} \alpha_k\right)}{\prod_{k=1}^{K} \Gamma\left(\alpha_k\right)} \prod_{i=1}^{M} \frac{\prod_{k=1}^{K} \Gamma\left(_i n_k^\star + \alpha_k\right)}{\Gamma\left(\sum_{k=1}^{K} {_i}n_k^\star + \alpha_k\right)} \prod_{j=1}^{{_i}n_\star^\star} \phi_{z_{ij}, w_{ij}} \qquad (3)$$

$$\propto \prod_{i=1}^{M} \left[\prod_{k=1}^{K} \Gamma\left(_i n_k^\star + \alpha_k\right)\right] \prod_{j=1}^{{_i}n_\star^\star} \phi_{z_{ij}, w_{ij}} \qquad (4)$$

To infer this distribution, we conduct posterior inference using Gibbs sampling, which requires the complete conditional distribution for any token topic assignment given values for all other variables in the model and the data (again, space considerations do not permit the full derivation here):

$$P\left(z_{mn} = t \mid \overline{\overline{z}}_{\neg mn}, \overline{\overline{w}}, \phi, \alpha\right) \propto \left(_m n_t^\star + \alpha_t\right) \phi_{t, w_{mn}} \qquad (5)$$

Because the topic-word counts in EDA are precomputed, the Gibbs sampler algorithm is somewhat simpler than LDA's. Pseudocode for the sampling algorithm is given in Algorithm 1.

Algorithm 1. Gibbs Sampler for Explicit Dirichlet Allocation

Input: *Word vector* $\overline{\overline{w}}$ *where* w_{ij} *is the index of the word type at position j of document i.*
Randomly initialize each element of $\overline{\overline{z}}$ *to values in* $\{1, 2, \ldots, K\}$
for $1 \leq i < M$ **do**
  for $1 \leq j < {_i}n_\star^\star$ **do**
    Sample $z_{ij} \sim UniformCategorical(K)$

Sample topic assignments from the complete conditional (eq. (5)) for N iterations
for N *iterations* **do**
  for $1 \leq i < M$ **do**
    for $1 \leq j < {_i}n_\star^\star$ **do**
      Sample $z_{ij} \sim P\left(z_{ij} = k \mid \overline{\overline{z}}_{\neg ij}, \overline{\overline{w}}\right)$

---

4   Experiments

In this section we present experiments, results, and analysis designed to assess the quality of the LDA-STWD and EDA topic models. We begin with a discussion of data, our implementation of the topic labeling algorithm described by Lau, Grieser, Newman, *et al.* [10], and our general approach to evaluating label quality using elicited human judgments.

Two datasets are used in the experiments: REUTERS 21578 and SOTU CHUNKS. REUTERS 21578 is a widely-used newswire dataset consisting of 11 367 news reports in 82 business-centric categories [12]. We tokenize the dataset's documents using MALLET's default stopword list, with the addition of the words "blah", "reuter", and "reuters", yielding a dataset of 827 841 tokens. Evaluations performed using this dataset also omit documents with fewer than 100 characters, fewer than 80 tokens, or more than 20% of characters being numeric.

The SOTU CHUNKS dataset was derived from the corpus of State of the Union messages delivered once a year (with minor variations) by United States presidents beginning with George Washington's first in 1790 and continuing to the present day. The messages are topically diverse due to the wide range of issues, times, and circumstances addressed. We split 223 publicly available State of the Union messages into 11 413 two-paragraph chunks to aid comprehension by human judges in the document label quality task.

We investigated the convergence properties of LDA-STWD and EDA by calculating log-likelihood of the data using the model at each iteration. On SOTU CHUNKS and REUTERS 21578, LDA-STWD sees rapid convergence, with the rate of change in log-likelihood dropping dramatically by the tenth iteration. This rapid convergence relative to LDA proper can be attributed to the lack of sampling of the topic-word distributions.

4.1 Baseline: Lau, et al.'s Post-Hoc Naming of LDA Topics

The topic labeling algorithm described in Lau, Grieser, Newman, *et al.* [10] is key to our evaluation strategy, because it represents the state of the art in automatic generation of topic labels. Since a working implementation of the algorithm was not readily available, it was necessary to write our own. Our implementation differs from the original by using Lucene rather than Google to index documents, using a more recent Wikipedia snapshot (3 April 2012), computing only pointwise mutual information rather than a variety of association measures, restricting fallback label candidates to words that are also Wikipedia article titles, and ignoring disambiguation pages rather than resolving them as the pages to which they point.

We conducted experiments to determine the best-performing topic count parameter for use in the LDA+LAU algorithm. Since LDA is parameterized by a number of topics K and LDA is a key component of LDA+LAU, it is necessary to set K. We do so by choosing the topic count whose labels are most preferred over labels produced with other topic counts. Topic count calibration user studies were performed for the SOTU CHUNKS and REUTERS 21578 datasets, using the same approach to topic labeling and document labeling tasks used as the main form of validation for LDA-STWD and EDA (see Section 4.2). In this case, LDA+LAU was compared to itself using different numbers of topics. Topic counts tried were 10, 20, 50, 100, and 200, using 5 distinct runs each. In the study, 3 users annotated 25 comparisons for each unordered pair of topic counts. The intention of the calibration is to allow LDA-STWD and EDA approaches to be compared to LDA+LAU at its best-performing setting, while not incurring the annotation cost required to perform the actual cross-method comparisons at all possible topic counts.

In the SOTU CHUNKS document labeling calibration study, the probability of participants choosing document labels produced by 10-topic LDA+LAU was 0.54, a higher rate of preference than for any other topic count. The result for REUTERS 21578 was nearly identical, with a preference probability of approx. 0.57, beating out all other topic counts evaluated. On the topic labeling task, SOTU CHUNKS was most preferred at 10 topics, with a probability of approx. 0.64. For REUTERS 21578, preference was maximized at 20 topics, with a preference probability of approx. 0.56.

4.2 Human Judgments of Label Quality

To assess LDA-STWD's and EDA's ability to find appropriate labeled topics given an input corpus, we compare document and topic labels from those algorithms to labels generated by the LDA+LAU combination. We run all three algorithms on the same target corpus. Their output is then used to create annotation tasks for Amazon Mechanical Turk to evaluate how well LDA-STWD or EDA labels fit their topics and documents, respectively. Amazon Mechanical Turk is a popular crowdsourcing platform that allows *requesters* to submit tasks for human *workers* to complete. Mechanical Turk has been shown to be an effective tool for a variety of data annotation tasks [16]. In particular it has been successfully used to evaluate the output of topic models [3].

Topic Label Quality. To assess LDA-STWD's and EDA's ability to find appropriate labeled topics given an input corpus, we compare topic labels from those algorithms to labels generated by (our implementation of) LDA+LAU. The $\langle topic, label \rangle$ pairs that LDA-STWD and EDA depend on are not products of the models but rather are inherent in the Wikipedia topic corpus. Thus any evaluation of topic label quality does not evaluate the quality of these models *per se*, but assesses a critical property of the topic corpus, relative to the quality of $\langle topic, label \rangle$ pairs generated by LDA+LAU.

In the topic label quality task, the user is presented with two $\langle topic, label \rangle$ pairs (one from each model) side-by-side and asked to choose the pair in which the label best matches the corresponding topic. Topics are represented using the top 10 words in the topic by $P(w \mid z)$. To make $\langle topic, label \rangle$ pairs from the two models more comparable, the topics taken from the topic corpus were restricted to the 100 most frequent topics in an EDA analysis of the relevant target corpus. While acknowledging the well-known weaknesses of traditional null hypothesis significance tests, we nevertheless employ the standard two-tailed binomial test in our analysis of results. Results of the topic label quality user study experiments are given in Table 2 and show the number of times users preferred document labels from the Wikipedia topic corpus over those generated by LDA+LAU. For example, various users were shown $\langle topic, label \rangle$ pair from the Wikipedia topic corpus alongside a $\langle topic, label \rangle$ pair generated by LDA+LAU a total of

Table 2. Outcome of topic label quality experiments

X	Y	X Preferred	Y Preferred	BT(X; Y, 0.5)
Wikipedia	LDA+LAU	201	207	0.805
Wikipedia	Random	64	31	0.000 924 6
LDA+LAU	Random	65	31	0.000 674 7

(a) On SOTU CHUNKS

X	Y	X Preferred	Y Preferred	BT(X; Y, 0.5)
Wikipedia	LDA+LAU	326	103	4.667×10^{-28}
Wikipedia	Random	77	34	5.495×10^{-5}
LDA+LAU	Random	30	30	1.0

(b) On REUTERS 21578

408 times, preferring the Wikipedia label 201 times and preferring the label generated by LDA+LAU 207 times. The last column in the table shows the outcome of a two-tailed binomial test, the probability that these results would be produced if users were equally likely to choose one label or the other ($p = 0.5$, i.e. the null hypothesis). The second and third rows show how each option fared against randomly generated labels, a "sanity check" against a naive baseline.

In the case of SOTU CHUNKS, $BT(201; 207, 0.5) \approx 0.805$, leaving no room to conclude any statisticaly significant difference between the two options. REUTERS 21578, on the other hand, displays a clear distinction: $BT(326; 103, 0.5) \approx 4.667 \times 10^{-28}$, meaning that we reject the null hypothesis with confidence at the $9.334 \times 10^{-26} \%$ level, concluding that participants clearly preferred the Wikipedia labels on this dataset.

Document Label Quality. In this section we describe experiments conducted to assess LDA-STWD and EDA in terms of performance on a document labeling task.

The user is presented with a randomly selected document from the target corpus. For each of the two topic models, the user is shown the labels of the top 10 topics in the document by $P(z \mid \theta)$ and asked to choose which of the two sets of labels best matches the content of the document.

In the document label quality task, participants were shown a short document and two possible labels for the document, one from LDA-STWD or EDA and one from LDA+LAU. They were then asked to choose which label best fit the document. To account for positional bias, in the prompt seen by participants the position (left or right) on screen was randomized. As an additional sanity check, occasionally one of the labels would be replaced with a randomly chosen label. Clearly, if the random labels were to outperform either of the models, then there would likely be something wrong with the experiment.

Table 3. Outcome of document label quality experiments on each dataset

X	Y	X Preferred	Y Preferred	BT(X; Y, 0.5)
LDA-STWD	LDA+Lau	256	160	2.905×10^{-6}
LDA-STWD	Random	54	27	0.003 596
LDA+Lau	Random	91	11	7.967×10^{-17}
EDA	LDA+Lau	123	266	3.286×10^{-13}
EDA	Random	82	26	6.141×10^{-8}
LDA+Lau	Random	91	11	7.967×10^{-17}

(a) SOTU Chunks

X	Y	X Preferred	Y Preferred	BT(X; Y, 0.5)
LDA-STWD	LDA+Lau	233	205	0.1970
LDA-STWD	Random	76	20	7.319×10^{-9}
LDA+Lau	Random	52	14	2.822×10^{-6}
EDA	LDA+Lau	182	247	0.001 97
EDA	Random	60	21	1.694×10^{-5}
LDA+Lau	Random	79	11	7.774×10^{-14}

(b) Reuters 21578

Results of the document label quality user study experiments for Latent Dirichlet Allocation with Static Topic-Word Distributions are given in Table 3.

In the case of SOTU Chunks, $BT(256; 160, 0.5) \approx 2.905 \times 10^{-6}$. We can thus firmly reject the null hypothesis, leaving us to conclude at the $5.810 \times 10^{-4} \%$ level that study participants preferred document labels generated by LDA-STWD over those generated by LDA+Lau on this dataset.

The results for Reuters 21578 are inconclusive: $BT(233; 205, 0.5) \approx 0.1970$, meaning the data are sufficiently well explained by the null model that there is not sufficient justification to conclude that either algorithm outperforms the other on this dataset.

Results of the document label quality user study experiments for Explicit Dirichlet Allocation in Table 3 reveal that EDA fared much worse than LDA-STWD. In the case of SOTU Chunks, $BT(123; 266, 0.5) \approx 3.286 \times 10^{-13}$, indicating that study participants preferred document labels generated by LDA+Lau over those generated by EDA on this dataset, with significance at the $6.571 \times 10^{-11} \%$ level. Likewise on Reuters 21578, $BT(182; 247, 0.5) \approx 0.001 97$, meaning that labels generated by LDA+Lau were preferred to those generated by EDA on this dataset, with significance at the 0.393% level.

5   Conclusions and Future Work

This paper introduces two methods for probabilistic explicit topic modeling that address specific weaknesses of LDA, namely the lack of useful topic labels and the inter-run identifiability problem. LDA-STWD does so by directly substituting precomputed counts for LDA topic-word counts, leveraging LDA's existing Gibbs sampler inference. EDA defines an entirely new explicit topic model and derives the inference method from first principles. Both of these methods approximate topic-word distributions *a priori* using word distributions from Wikipedia articles, with each article corresponding to one topic and the article title being used as a topic label.

LDA-STWD significantly outperforms LDA+LAU in labeling the documents in the SOTU CHUNKS corpus, and the topic labels derived from Wikipedia are vastly preferred by human annotators over those generated by LDA+LAU for the REUTERS 21578 corpus. A number of non-rejections of the null hypothesis also speak in favor of LDA-STWD and EDA as more-principled peers to LDA+LAU. The lack of significant difference between Wikipedia and LDA+LAU on SOTU CHUNKS shows the Wikipedia-derived topics performing no worse than the state of the art. Likewise, LDA-STWD is not found to be worse than incumbent LDA+LAU on REUTERS 21578, even though it cannot be said to be significantly better. And finally, EDA is not found to perform significantly worse than LDA+LAU on REUTERS 21578. Thus the principled, straightforward LDA-STWD and EDA algorithms can be seen performing at the same level of quality as the less-intuitive *post hoc* Lau, et al. method for this task.

The superiority of LDA-STWD to EDA demands explanation. Mathematically, the two models differ solely in the presence or absence of smoothing in the topic-word distributions (ϕ). This suggests that EDA's lack of smoothing could be harming performance. A variation on EDA worth investigating is a model in which topic-word distributions are latent, but explicit topic information is encoded in the priors on those distributions. Such an approach was taken in MR. LDA, but was not applied to a large number of topics [18].

Other avenues for future work include allowing some topics in LDA-STWD or EDA to have latent topic-word distributions, and seeking increased runtime performance through further parallelization of the algorithms.

References

[1] Blei, D.M.: Introduction to probabilistic topic models. Comm. of ACM (2011)
[2] Blei, D.M., Ng, A.Y., Jordan, M.I.: Latent Dirichlet Allocation. J. of Machine Learning Research 3, 993–1022 (2003)
[3] Chang, J., Boyd-Graber, J., Gerrish, S., Wang, C., Blei Reading, D.: tea leaves: How humans interpret topic models. In: NIPS (2009)
[4] Cimiano, P., Schultz, A., Sizov, S., Sorg, P., Staab, S.: Explicit versus latent concept models for cross-language information retrieval, pp. 1513–1518 (2009)
[5] Deerwester, S., Dumais, S., Furnas, G., Landauer, T., Harshman, R.: Indexing by Latent Semantic Analysis. J. Amer. Soc. Inf. Sci. 41(6), 391–407 (1990)
[6] Gabrilovich, E., Markovitch, S.: Computing semantic relatedness using Wikipedia-based Explicit Semantic Analysis. IJCAI 6, 12 (2007)

[7] Gardner, M.J., Lutes, J., Lund, J., Hansen, J., Walker, D., Ringger, E., Seppi, K.: The Topic Browser: An interactive tool for browsing topic models. NIPS Workshop on Challenges of Data Visualization (2010)

[8] Griffiths, T.L., Steyvers, M.: Finding scientific topics. PNAS 101(suppl. 1), 5228–5235 (2004)

[9] Hofmann, T.: Unsupervised learning by Probabilistic Latent Semantic Analysis. Machine Learning 42(1), 177–196 (2001)

[10] Lau, J.H., Grieser, K., Newman, D., Baldwin, T.: Automatic labelling of topic models. HLT 1, 1536–1545 (2011)

[11] Lau, J.H., Newman, D., Karimi, S., Baldwin, T.: Best topic word selection for topic labelling. In: COLING, pp. 605–613 (2010)

[12] Lewis, D.D.: An evaluation of phrasal and clustered representations on a text categorization task. In: SIGIR, pp. 37–50 (1992)

[13] Magatti, D., Calegari, S., Ciucci, D., Stella, F.: Automatic labeling of topics. In: ISDA, pp. 1227–1232 (2009)

[14] Mei, Q., Shen, X., Zhai, C.X.: Automatic labeling of multinomial topic models. In: KDD, pp. 490–499 (2007)

[15] Ramage, D., Hall, D., Nallapati, R., Manning, C.: Labeled LDA: A supervised topic model for credit attribution in multi-labeled corpora. In: EMNLP, vol. 1, pp. 248–256 (2009)

[16] Snow, R., O'Connor, B., Jurafsky, D., Ng, A.: Cheap and fast—but is it good?: Evaluating non-expert annotations for natural language tasks. In: EMNLP, pp. 254–263 (2008)

[17] Spitkovsky, V.I., Chang, A.X.: A cross-lingual dictionary for english Wikipedia concepts. In: LREC, May 23-25 (2012)

[18] Zhai, K., Boyd-Graber, J., Asadi, N., Alkhouja, M.: Mr. LDA: A flexible large scale topic modeling package using variational inference in map/reduce. In: WWW (2012)

Decision Tree-Based Evaluation of Genitive Classification – An Empirical Study on CMC and Text Corpora

Sandra Hansen and Roman Schneider

Institute for German Language (IDS), Mannheim/Germany
{hansen,schneider}@ids-mannheim.de

Abstract. Contemporary studies on the characteristics of natural language benefit enormously from the increasing amount of linguistic corpora. Aside from text and speech corpora, corpora of computer-mediated communication (CMC) position themselves between orality and literacy, and beyond that provide insight into the impact of "new", mainly internet-based media on language behaviour. In this paper, we present an empirical attempt to work with annotated CMC corpora for the explanation of linguistic phenomena. In concrete terms, we implement machine learning algorithms to produce decision trees that reveal rules and tendencies about the use of genitive markers in German.

Keywords: Corpus Linguistics, Computer-Mediated Communication, Machine Learning, Decision Trees, Grammar, Genitive Classification.

1   Introduction

Linguistic studies are increasingly corpus-based, i.e. their statements rely on empirical data, computed on the basis of natural language. Due to the problematic nature of speech corpora, e.g. the difficulty of achieving substantial amounts of authentical spoken samples or the influence of situational conditions on the proband's speech behavior, text corpora currently represent the vast majority of available resources. In this situation, corpora of computer-mediated communication (CMC) open up new possibilites for the examination of language phenomena between the poles of orality and literacy [2] [5]. Internet-based discourse genres such as e-mails, weblogs, or chat and discussion groups offer insight into the use of language in situations that are at least to some extent close to verbal data and face-to-face communication [7].

It is well known that, due to specific production conditions, the syntactical rules of spoken and computer-mediated language differ from the rules that apply to written language. This has substantial impact on the performance of linguistic tools like taggers and parsers. Therefore it seems most desirable to verify the conditions under which automatically annotated CMC corpora can contribute to linguistic research. We are especially interested in the question whether the statistical evaluation of hypotheses based on machine learning algorithms is applicable. For a first estimate, we compare the results of an empirical study conducted on the basis of a large text corpus with the output of the same methods and algorithms adapted to CMC data.

I. Gurevych, C. Biemann, and T. Zesch (Eds.): GSCL 2013, LNAI 8105, pp. 83–88, 2013.
© Springer-Verlag Berlin Heidelberg 2013

The evaluation of hypotheses predicting the use of genitive markers in German is a field of study that is notoriously complicated and generates cases of doubt, because there is no generally accepted model: Is it better to use "des Films" or "des Filmes" (i.e., to use "-s" or "-es" marker)? Under which conditions is it tolerable to omit the genitive marker (e.g., zero-marker as in "des Internet")? In order to find an empirical answer, manifold intra- and extralinguistic parameters has to be considered: the number of word syllables, types of coda, noun frequency, information about medium, register, and region etc. Therefore, decision trees seems to be a valuable tool to identify, order, and structure the factors that are most prominent for the actual decision.

2   Corpus Resources

For our study, we used the *Dortmund Chat Corpus*[1] that was compiled between 2003 and 2009. It covers logfiles from different chat groups, supplemented with CMC-specific metadata and encoded in an interchangeable XML format, ranging over a variety of subjects and situational contexts. Though the complete corpus contains more than one million word forms, the publicly available release has to content itself with 548,067 word forms within 59,558 chat postings. For our further processing, the original chat texts were annotated morphosyntactically with three competing systems: Connexor Machinese Tagger, TreeTagger, and Xerox Incremental Parser[2]. In the following, we primarily use the Xerox parser because it gives us the broadest range of syntactic and structural annotation, for example case information for nouns. As text-oriented counterpart, we choose the 2011-I release of the *German Reference Corpus DeReKo*[3] with more than 4 billion word forms, which is one of the major resources worldwide for the study of written German. Like for the CMC corpus, morphosyntactic annotations from the three tools mentioned above are added, and the corpus is enriched with a comprehensive set of extra-linguistic metadata. Language samples, annotations, and metadata were integrated into a prototypical RDBMS-driven corpus storage and retrieval framework. This system allows for the flexible analysis of multi-layered corpora with regular expressions and a combined search on all available types of annotation and metadata, using parallelized SQL queries and a MapReduce-like retrieval paradigma. Our study benefits from the fact that within the framework all language samples are stored wordwise, and every wordform is connected to intra- and extra-linguistic metadata according to an efficient logical data model [9].

3   The Genitive Extraction

The primary corpus data served as a basis to extract all relevant genitive forms. As a first step, the genitive candidates were filtered out using a specifically adjusted Perl script. The resulting database consisted of about 454,500 types and 7,334,500 tokens.

---

[1] http://www.chatkorpus.tu-dortmund.de

[2] See http://www.connexor.eu/technology/machinese/index.html, http://www.cis.uni-muenchen.de/~schmid/tools/TreeTagger/, and http://open.xerox.com/Services/XIPParser/, respectively.

[3] http://www.ids-mannheim.de/DeReKo/

Then, by means of the script and in order to weight the findings, several distribution rules were checked. For example, in cases where the word ends with a genitive marker and the lemma does not end with a marker, the genitive candidate gets a so-called score point. In the context of a second distribution rule, we give an additional score point for a typical pre- or post-modified genitive preposition. If the script detects an adjacent genitive article or a genitive article within a certain distance from a premodificated, adjacent adjective, it assumes the presence of a noun with a genitive form, and the token in question gets two more score points. The following example shows a genitive noun (token = "*Anblicks*"; lemma = "*anblick*") with a genitive preposition ("*wegen*") followed by a genitive article ("*des*") and a premodificated adjective ("*schönen*"): "*wegen des schönen Anblicks*".

Overall, we implemented 19 different distribution rules, and counted the total of the assigned score points for every genitive candidate. The higher the score points, the more likely the candidate was considered a genitive noun. All candidates with score points greater than two were taken into account. The script output was measured against a manually annotated gold standard, containing 9,000 nouns extracted out of 1,000 sentences. Precision, recall, and F-scores are about 95%.

Within a following step, the candidates were enriched with metadata (location, medium, domain, year etc.) and morphosyntactic information in order to get additional grammatical evidence (e.g., information about the genus). We isolated loanwords, acronyms, and neologisms using existing word lists. Some distributionally motivated information was extracted with a second Perl script. By comparing our data set with CELEX [1], we were able to include phonetic and prosodic information (e.g., the number of syllables or the character of the last sound/coda) into our calculations.

4   Statistical Analysis

Subsequently, we evaluated the factors influencing the use of genitive markers, and tried to model a decision tree for both corpora based on the token data. We started by encoding 34 language-immanent and extra-linguistic factors influencing the marking of a genitive noun. To get a general idea about the specific factor's influences and side effects, we calculated chi-square-tests and visualized the residuals with an association plot (cf. [3] [4] [6])[4]. The plots visualize the standard deviations of the observed frequencies as a function of the expected frequencies. Each cell is represented by a rectangle, whose height is proportional to the residual of the cell, and having a width proportional to the square root of the expected frequency. Therefore, the area of the rectangle is proportional to the difference between observed and expected frequencies.

As an in-depth presentation of all factors would exceed the limits of this paper, we will concentrate on a rather small selection. Figure 1 represents the influence of the number of syllables to the genitive marker in the DeReKo-based text corpus. The association plot shows the under-representation of the "ns"-, zero-, "s"- and "ses"-markers and the over-representation of the "ens"- and "es"- markers as a function of the number of syllables. If the lexeme consists of multi syllables, the "ens"- and "es"-markers are under-represented and the residuals of the "s"-markers are much higher.

---

[4] All tests and plots were conducted and produced using the VCD package (Visualizing Categorial Data) of the statistical software "R".

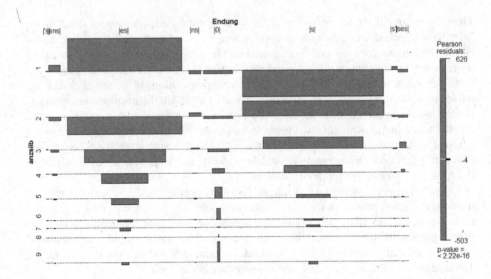

Fig. 1. Association plot for the influence of number of syllables (text corpus)

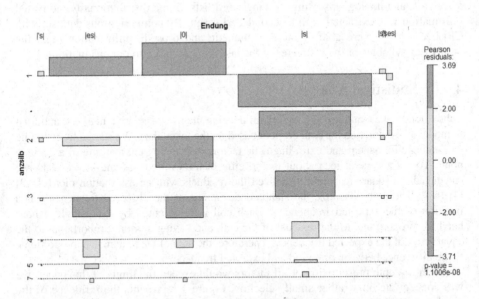

Fig. 2. Association plot for the influence of number of syllables (CMC corpus)

Figure 2 shows the same statistical analysis on the CMC data set. Concerning the "es"- and "s"-markers, the results are similar to those of the text corpus. But interestingly, the residuals of the zero-marker (value "0" in the plots above) show different trends. Within the chat corpus, the zero-marker of words with one syllable is strongly over-represented, whereas the zero-marker in multi syllable words is significantly under-represented. Here, some in-depth linguistic interpretation would be valuable.

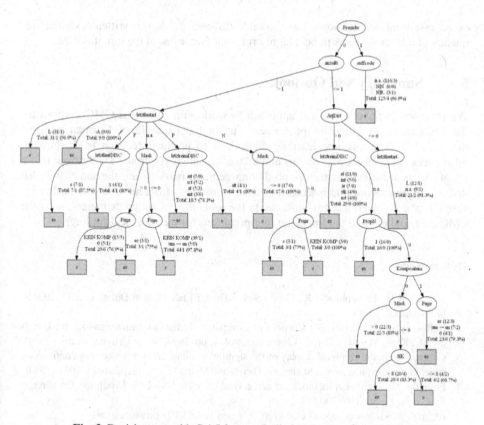

Fig. 3. Decision tree with C 4.5 factors predicting the use of genitive markers

The association plots and chi-square tests were produced and conducted for every single factor in order to test their influence on the distribution of genitive markers.

Afterwards, the main analysis was focused on the calculation and visualization of complex decision trees for each corpus. Decision tree learning uses tree representations as a predictive model that maps observations about an item to conclusions about the item's target value. The leaves of the trees represent class labels, the branches represent conjunctions of features that lead to these class labels.

For the statistical analysis, we took recourse to the C4.5 algorithm [8] using the WEKA software [10]. We calculated decision trees for both the chat corpus and the text corpus in order to compare the various relationships of the factors predicting the use of genitive markers. As a result, 91.35% of the instances in the text corpus were correctly classified. In the CMC corpus, the correctly classified instances are 90.89%.

Figure 3 displays the decision tree for the prediction of genitive markers in the chat corpus. It has to be read as follows: The first significant single factor splitting the data into separate groups is whether the noun is – or is not – a loanword (leaf "Fremdw"). Then, if the lexeme is part of the basic vocabulary (branch "0"), the factor "number of syllables" (leaf "anzsilb") is relevant, and so on. The complete decision tree for the text corpus is far too large and comprehensive to be included in this paper, but it can be accessed online.[5] It shows some notable differences, as for written texts the frequency of a lexeme seems to be a highly relevant factor just at the top of the tree.

5   Summary and Outlook

We presented a novel empirical approach to work with annotated CMC corpora for the explanation of linguistic phenomena, using the example of German genitive markers. We used machine learning algorithms to produce decision trees showing differences between a CMC corpus and a "traditional" text corpus. They reveal that a lot of the most influential factors predicting genitive marking are the same, but also that the sequences and the interaction of the factors are different. We will further investigate the involved mechanisms, verify the reliability of automated annotations for CMC data, and explore the linguistic interpretations for the statistical findings.

References

1. Baayen, R.H., Piepenbrock, R., Gulikers, L.: The CELEX Lexical Database (CD- ROM), Philadelphia (1995)
2. Beißwenger, M., Storrer, A.: Corpora of Computer-Mediated Communication. In: Lüdeling, A., Kytö, M. (eds.) Corpus Linguistics, vol. 1, pp. 292–308. de Gruyter, Berlin (2008)
3. Cohen, A.: On the graphical display of the significant components in a two-way contingency table. In: Communications in Statistics - Theory and Methods, vol. A9, pp. 1025–1041 (1980)
4. Friendly, M.: Graphical methods for categorical data. In: SAS User Group Int. Conference Proc., vol. 17, pp. 190–200 (1992),
   http://www.math.yorku.ca/SCS/sugi/sugi17-paper.html
5. Herring, S.: Computer-Mediated Conversation. Language@Internet 7/8 (2010/2011),
   http://www.languageatinternet.org
6. Meyer, D., Zeileis, A., Hornik, K.: The strucplot framework: Visualizing multi-way contingency tables with vcd. Report 22, Department of Statistics and Mathematics, Wirtschaftsuniversität Wien, Research Report Series (2005)
7. Ogura, K., Nishimoto, K.: Is a Face-to-Face Conversation Model Applicable to Chat Conversations? In: Proc. PRICAI Workshop Language Sense on Computer, pp. 26–31 (2004)
8. Quinlan, J.R.: C4.5: Programs for Machine Learning. Morgan Kaufmann, San Francisco (1993)
9. Schneider, R.: Evaluating DBMS-Based Access Strategies to Very Large Multi-Layer Annotated Corpora. In: Proceedings of the LREC-2012 Workshop on Challenges in the Management of Large Corpora, Istanbul (2012)
10. Witten, I., Frank, E.: Data Mining: Practical Machine Learning Tools and Techniques. Morgan Kaufmann, San Francisco (2005)

---

[5] The tree can be accessed here:
   http://hypermedia.ids-mannheim.de/treeText.c095.m2000.pdf

Extending the TüBa-D/Z Treebank with GermaNet Sense Annotation

Verena Henrich and Erhard Hinrichs

University of Tübingen, Department of Linguistics,
Wilhelmstr. 19, 72074 Tübingen, Germany
{vhenrich,eh}@sfs.uni-tuebingen.de
http://www.sfs.uni-tuebingen.de

Abstract. This paper describes the manual construction of a sense-annotated corpus for German with the goal of providing a gold standard for word sense disambiguation. The underlying textual resource, the TüBa-D/Z treebank, is a German newspaper corpus already manually enriched with high-quality, manual annotations at various levels of grammar. The sense inventory used for tagging word senses is taken from GermaNet [8,9], the German counterpart of the Princeton WordNet for English [6]. With the sense annotation for a selected set of 109 words (30 nouns and 79 verbs) occurring together more than 15 500 times in the TüBa-D/Z, the treebank currently represents the largest manually sense-annotated corpus available for GermaNet.

Keywords: Sense-annotated corpus, sense-tagged corpus, GermaNet, TüBa-D/Z treebank.

1 Introduction

Sense-annotated corpora are a prerequisite for (semi-)supervised approaches to a wide variety of natural language processing tasks, including word sense disambiguation (WSD), statistical machine translation and other semantics-related tasks. It is therefore not surprising that most research on WSD has focused on languages such as English for which sense-annotated corpora have been available [1,5,12] and considerably less on languages with a shortage of such corpora.

The purpose of the present paper is to help close this gap by describing the manual construction of a sense-annotated corpus for German, a language for which the availability of sense-annotated corpora is restricted. Thus far there is only one sense-annotated corpus freely available for GermaNet, which is called WebCAGe (short for: *Web-Harvested Corpus Annotated with GermaNet Senses*) [11]. WebCAGe has been constructed semi-automatically by annotating web-harvested texts with senses from GermaNet. Unfortunately, this corpus has several limitations: (i) only parts of WebCAGe are freely available due to legal restrictions on some of the underlying textual materials, (ii) the underlying textual materials are web-harvested and thus depend on the varying quality of web texts, and (iii) due to the harvesting method, the number of annotated target

I. Gurevych, C. Biemann, and T. Zesch (Eds.): GSCL 2013, LNAI 8105, pp. 89–96, 2013.

word occurrences is skewed and cannot be used to compile most frequent sense information. In order to overcome these shortcomings, the TüBa-D/Z treebank is extended by sense annotation for a selected set of lemmas. The use of treebank data is motivated by the following considerations:

1. The grammatical information contained in a treebank makes it possible to utilize a much richer feature set for automatic WSD compared to sense-annotated training data that otherwise contain little or no linguistic annotation [7,3]. This is particularly useful for automatic WSD of verbs where the syntactic structure in which a verb occurs is often highly predictive of different word senses.
2. Since the TüBa-D/Z is based on a newspaper corpus, this ensures a broad coverage of topical materials such as politics, economy, society, environmental issues, sports, arts and entertainment. This broad coverage of topics also makes it possible to obtain reliable information about the relative frequency of different senses of a given word.

The remainder of this paper is structured as follows: Section 2 provides a brief overview of the resources GermaNet and TüBa-D/Z. The selection of words to be manually sense-annotated is elaborated in Section 3. The annotation process as well as inter-annotator agreement are reported in Section 4. Finally, related work is discussed in Section 5, together with concluding remarks.

2   Resources

The Sense Inventory for the sense-annotated corpus presented here is taken from GermaNet [8,9], a lexical semantic network that is modeled after the Princeton WordNet for English [6]. It represents semantic concepts as *synsets*, i.e., as sets of (near-)synonymous words (referred to as *lexical units*), that are interlinked by semantic relations. GermaNet covers the three word categories of adjectives, nouns, and verbs, each of which is hierarchically structured in terms of the hypernymy relation of synsets. GermaNet's version 8.0 (release of April 2013) contains 111 361 lexical units, which are grouped into 84 584 synsets. Using a wordnet as the gold standard for the sense inventory is fully in line with standard practice for English where the Princeton WordNet is typically taken.

The Underlying Textual Resource for the sense-tagged corpus is the syntactically annotated Tübingen Treebank of Written German (TüBa-D/Z) [17], the largest manually annotated treebank for German. It includes the following annotation layers: part-of-speech (using the Stuttgart-Tübingen tagset (STTS) [16]), inflectional morphology, lemmatization, syntactic constituency, grammatical functions, named entity classification, anaphora and coreference relations. The textual material for the treebank is taken from the daily newspaper "die tageszeitung" (taz). Each article is split into paragraphs and sentences, and each sentence in turn into tokens. Each token is enriched with its lemma, its STTS part-of-speech tag, and its inflectional morphology such as case, gender, number, person, etc. The current release 8 of the TüBa-D/Z contains 1 365 642 tokens occurring in 75 408 sentences that are taken from 3 256 newspaper articles.

3   Selection of Words to Be Sense-Annotated

The sense annotation in the TüBa-D/Z is geared toward the lexical sample task in WSD, rather than toward the all-words task. The decision against sense annotation of all words of running text in a selected subcorpus is motivated by the requirements of machine learning as the intended use of the data. Such data are useful for training (semi-)automatic machine learning models only if there are sufficiently many instances for each item to be classified. Due to limitations of how much text can reasonably be annotated manually in an all-words, sense-annotated corpus, the resulting numbers of instances for each token are not of sufficient frequency for machine-learning applications. The selection of lemmas to be sense-annotated in the TüBa-D/Z was guided by the following criteria:

1. The selected lemmas have at least two senses in GermaNet and occur at least 16 times in the TüBa-D/Z.
2. The sample as a whole represents a good balance of frequencies and number of distinct word senses.
3. The selected words include both nouns and verbs so as to be able to compare and evaluate the efficacy of structured linguistic information present in treebanks across the two word classes.
4. For verbs, the selected lemmas display different degrees of correlations between differences in word senses and valence frames.

As a result of the above criteria, a total of 30 lemmas for nouns and 79 lemmas for verbs were selected. Table 1 provides an overview of the entire lexical sample annotated in the TüBa-D/Z.

Table 1. Quantitative statistics of sense-annotated words

	Nouns	Verbs
Total # of annotated word lemmas	30	79
Total # of occurrences in TüBa-D/Z	7 538	7 967
Frequency range (occurrences/lemma)	22-1427	16-710
Average frequency (occurrences/lemma)	251	101
Polysemy range (senses/lemma)	2-7	2-14
Average polysemy (senses/lemma)	3.97	2.84

Altogether, 7 967 verb occurrences are annotated with the senses of 79 verb lemmas (see Table 1). The average occurrence per verb lemma is 101 with the least frequent verb occurring 16 times, the most frequent one 710 times. The average polysemy (number of senses in GermaNet) is 2.84, with the most polysemous verb showing 14 senses in GermaNet.

While most of the selection criteria described above are self-evident, the selection of verbs, which is guided by different degrees of correlations among word senses and valence frames, deserves further explanation. For verbs, the syntactic structure in which a verb occurs is often highly predictive of different word senses. The German verb *enthalten* is a case in point. It has two distinct word senses of 'contain' and 'abstain', which correspond directly to two distinct valence frames. The former requires a valence frame with a nominative and with an accusative object as in *Das Medikament enthält Alkohol* ('The medicine contains alcohol'), while the latter requires a reflexive pronoun as its object as in *Ich enthalte mich eines Urteils* ('I abstain from passing judgment').

Verbs differ, however, in the degree of correlation between word senses and valence frames, ranging from total correlation to complete lack of correlation. The German verb *begrüßen* is an example of the latter kind. Its senses of 'greet someone' and 'have a positive attitude toward something' both have a valence frame with a nominative and accusative noun phrase.

The inclusion of different degrees of correlations between word senses and valence frames for verbs to be sense-annotated in TüBa-D/Z makes it possible to systematically assess the impact of information about syntax (e.g. valence) and of lexical semantics (e.g. the collocational behavior of the target lemma) in machine-learning models for WSD. Ideally, features encoding valence information should suffice for verbs such as *enthalten*, while for WSD of verbs such as *begrüßen* they carry no weight whatsoever. In order to provide a fine-grained spectrum of possible degrees of correlations between word senses and valence frames for verbs, the verbs selected for sense annotation fall into four distinct classes:

- **Class 1:** All verbs in this class have distinct valence frames for their word senses, such as the above-illustrated example of *enthalten*.
- **Class 2:** This class contains verbs where at least one sense has a distinct valence frame. For example, the two senses 'withdraw' and 'extract' of *entziehen* both have the same valence frame of a nominative, an accusative, and a dative object, whereas the third sense 'to shirk doing sth.' requires a nominative and a dative object as well as a reflexive pronoun.
- **Class 3:** All verbs in this class share the same frame valence for all of their senses, as the above given example of *begrüßen* where both senses have a valence frame with a nominative and accusative noun phrase.
- **Class 4:** This class comprises verbs that do not fall into any of the classes 1-3. It contains for example the verb *liefern* with altogether four senses. Two senses share the same nominative/accusative valence frame and the other two senses have an additional dative object. Hence this verb does not belong to any of the classes 1-3.

The valence information used for assigning the verb lemmas to the four classes described above is taken from GermaNet, which uses the CELEX encoding of subcategorization frames within the lexical entries of verbs.

In comparison to the selection of the verbs to be sense-annotated, the selection of nouns seems rather simple. They are chosen by frequency and polysemy so as to be able to analyze the impact of different amounts of instances in the training data and different degrees of polysemy on automatic WSD. In total, 30 noun lemmas have been selected for manual sense annotation. Altogether, these nouns occur 7 538 times - at least 22 times and at most 1 427 times (see Table 1). On average, there are 251 occurrences per noun lemma. The average polysemy is 3.97 for the annotated nouns, ranging from 2-7 senses.

4 Annotation Process and Inter-Annotator Agreement

In order to assure good quality of the manual sense annotation and to be able to calculate inter-annotator agreement (IAA), the sense annotation is independently performed by two annotators (native German computational linguists) for all word lemmas and occurrences. The annotators have the possibility to indicate problematic word occurrences with comments to be discussed separately. The supervision of the two annotators is conducted by an experienced lexicographer, who is a native speaker of German and who has been the main responsible expert for the lexicographic extension of GermaNet for several years. If the TüBa-D/Z contains word senses of the selected lemmas that are currently not covered by GermaNet, the lexicographic expert decides whether to add those senses to GermaNet. In an adjudication step, the expert goes through all occurrences, where the two annotators either do not agree or at least one of them had a comment, and resolves disagreements.

Inter-annotator agreement is calculated to assess the reliability of the manual sense annotations. The Dice coefficient[1] is 96.4% for the 7 538 noun occurrences and 93.7% for the 7 967 verb occurrences, corresponding to Cohen's Kappa [4] values of 85.4% and 82.4%, respectively.[2] Overall, the percentage of IAA is very high. The values are comparable to the agreement statistics reported by Raileanu et al. [15] for their work on creating a German sense-annotated corpus.

The observed agreement values are much higher than those observed for English. Véronis [18], for example, observes a pairwise Dice coefficient of 73% for nouns and 63% for verbs. Palmer et al. [14] report an inter-annotator agreement of 71.3% for the English verb lexical sample task for SensEval-2. The reason of much higher IAA values for German than for English is obviously the number of distinct senses: an average of 3.97 for German nouns and 2.84 for German verbs (see column 6 in Table 1) as opposed to an average of 7.6 for English nouns and 12.6 for English nouns in the case of Véronis [18, Table 3].

The detrimental effect of very fine-grained word senses was already observed by Palmer et al. [14, p. 97]. They report an improvement in the inter-annotator agreement from 71.3 to 82% for the same SensEval-2 lexical sample task when more coarse-grained verb senses are used instead of the fine-grained distinctions taken from WordNet 1.7.

---

[1] The reported values are obtained by averaging the Dice coefficient for all annotated occurrences of the word category in question.

[2] Since Cohen's Kappa does not allow multiple categories (i.e., multiple senses) for a word, we follow Raileanu et al.'s technique [15] of ignoring all words where one of the two annotators selected more than one sense – in this study, 60 for nouns and 17 for verbs. Furthermore, when both annotators always pick one sense for all occurrences of a lemma, the Kappa coefficient is not informative and those occurrences are ignored in calculating the reported average.

A detailed inspection of the IAA for single words did not show a correlation between the IAA and the number of senses a word has. This finding corroborates the results reported by Fellbaum et al. [7] on sense-annotating the English Penn Treebank [13] with senses from the Princeton WordNet.

5   Related and Future Work

All sense-annotated corpora that have been created for research on word sense disambiguation for German have served the lexical sample task. Broscheit et al. [2] annotate the GermaNet senses of 40 word lemmas (6 adjectives, 18 nouns, and 16 verbs) in more than 800 occurrences in the deWAC corpus. Raileanu et al. [15] annotate 2 421 occurrences of the EuroWordNet-GermaNet senses of 25 nouns in a medical corpus obtained from scientific abstracts from the Springer Link website. The same medical corpus is annotated by Widdows et al. [19] with 24 ambiguous UMLS types – each of which occurs at least 11 times in the corpus (seven occur more than 100 times). Henrich et al. [11] constructed the sense-annotated corpus WebCAGe semi-automatically by annotating more than 10 000 word occurrences in web-harvested texts with GermaNet senses of more than 2 000 word lemmas. WikiCAGe (Henrich et al. [10]) contains more than 1 000 word lemmas semi-automatically annotated with GermaNet senses in more than 24 000 word occurrences in Wikipedia articles.

In terms of quantity, the present study significantly goes beyond previous efforts on manually creating a German sense-annotated corpus in that more than 15 500 occurrences from the TüBa-D/Z treebank are annotated with the GermaNet senses of 30 nouns and 79 verbs. The sense annotations will be made freely available for academic research as part of future releases of the treebank.[3]

What sets the presented work apart from related work is the systematic selection of verb lemmas to be manually sense-annotated. This selection is determined by the goal to analyze the influence of syntax and semantic on automatic WSD which is particularly interesting for verbs where the syntactic structure in which a verb occurs is often highly predictive of different word senses. The implementation of automatic WSD using contextual features and to investigate on the influence of syntax and semantic on automatic WSD is future work.

Acknowledgments. The research reported in this paper was jointly funded by the SFB 833 grant of the DFG and by the CLARIN-D grant of the BMBF. We are very grateful to Valentin Deyringer, Reinhild Barkey, and Christina Hoppermann for their help with the annotations reported in this paper. Special thanks go to Yannick Versley for his support with the sense-annotation tool and to Scott Martens and Christian M. Meyer for their input on how to calculate inter-annotator agreement.

---

[3] http://www.sfs.uni-tuebingen.de/en/ascl/resources/corpora/
sense-annotated-tueba-dz.html

References

1. Agirre, E., Marquez, L., Wicentowski, R.: Proceedings of the 4th International Workshop on Semantic Evaluations. Association for Computational Linguistics, Stroudsburg (2007)
2. Broscheit, S., Frank, A., Jehle, D., Ponzetto, S.P., Rehl, D., Summa, A., Suttner, K., Vola, S.: Rapid bootstrapping of Word Sense Disambiguation resources for German. In: Proceedings of the 10. Konferenz zur Verarbeitung Natürlicher Sprache, Saarbrücken, Germany, pp. 19–27 (2010)
3. Chen, J., Palmer, M.: Improving English Verb Sense Disambiguation Performance with Linguistically Motivated Features and Clear Sense Distinction Boundaries. In: Language Resources and Evaluation, vol. 43, pp. 181–208. Springer, Netherland (2009)
4. Cohen, J.: A Coefficient of Agreement for Nominal Scales. Educational and Psychological Measurement 20(1), 37–46 (1960)
5. Erk, K., Strapparava, C.: Proceedings of the 5th International Workshop on Semantic Evaluation. Association for Computational Linguistics, Stroudsburg (2010)
6. Fellbaum, C. (ed.): WordNet – An Electronic Lexical Database. The MIT Press (1998)
7. Fellbaum, C., Palmer, M., Dang, H.T., Delfs, L., Wolf, S.: Manual and Automatic Semantic Annotation with WordNet. In: SIGLEX Workshop on WordNet and other Lexical Resources, NAACL 2001, Invited Talk, Pittsburgh, PA (2001)
8. Hamp, B., Feldweg, H.: GermaNet – a Lexical-Semantic Net for German. In: Proceedings of ACL Workshop Automatic Information Extraction and Building of Lexical Semantic Resources for NLP Applications, Madrid (1997)
9. Henrich, V., Hinrichs, E.: GernEdiT – The GermaNet Editing Tool. In: Proceedings of the Seventh Conference on International Language Resources and Evaluation (LREC 2010), Valletta, Malta, pp. 2228–2235 (2010)
10. Henrich, V., Hinrichs, E., Suttner, K.: Automatically Linking GermaNet to Wikipedia for Harvesting Corpus Examples for GermaNet Senses. Journal for Language Technology and Computational Linguistics (JLCL) 27(1), 1–19 (2012)
11. Henrich, V., Hinrichs, E., Vodolazova, T.: WebCAGe - A Web-Harvested Corpus Annotated with GermaNet Senses. In: Proceedings of the 13th Conference of the European Chapter of the Association for Computational Linguistics (EACL 2012), Avignon, France, pp. 387–396 (2012)
12. Mihalcea, R., Chklovski, T., Kilgarriff, A.: Proceedings of Senseval-3: Third International Workshop on the Evaluation of Systems for the Semantic Analysis of Text, Barcelona, Spain (2004)
13. Marcus, M., Santorini, B., Marcinkiewicz, M.: Building a large annotated corpus of English: The Penn Treebank. In: Computational Linguistics, vol. 19, pp. 313–330 (1993)
14. Palmer, M., Ng, H.T., Dang, H.T.: Evaluation of WSD Systems. In: Agirre, E., Edmonds, P. (eds.) Word Sense Disambiguation: Algorithms and Applications, pp. 75–106. Springer (2006)
15. Raileanu, D., Buitelaar, P., Vintar, S., Bay, J.: Evaluation Corpora for Sense Disambiguation in the Medical Domain. In: Proceedings of the 3rd International Language Resources and Evaluation (LREC 2002), Las Palmas, Canary Islands, pp. 609–612 (2002)
16. Schiller, A., Teufel, S., Thielen, C.: Guidelines für das Tagging deutscher Textcorpora mit STTS. Technical report, Universities of Stuttgart and Tübingen (1995)

17. Telljohann, H., Hinrichs, E.W., Kübler, S., Zinsmeister, H., Beck, K.: Stylebook for the Tübingen Treebank of Written German (TüBa-D/Z). Technical report, Department of General and Computational Linguistics, University of Tübingen, Germany (2012)
18. Véronis, J.: A study of polysemy judgments and inter-annotator agreement. In: Proceedings of SENSEVAL-1, Herstmonceux Castle, England (1998)
19. Widdows, D., Peters, S., Cederberg, S., Chan, C.-K., Steffen, D., Buitelaar, P.: Unsupervised monolingual and bilingual word-sense disambiguation of medical documents using umls. In: Proceedings of the ACL 2003 Workshop on Natural Language Processing in Biomedicine, BioMed 2003, pp. 9–16. Association for Computational Linguistics, Stroudsburg (2003)

Topic Modeling for Word Sense Induction

Johannes Knopp, Johanna Völker*, and Simone Paolo Ponzetto

Data & Web Science Research Group
University of Mannheim, Germany
{johannes,johanna,simone}@informatik.uni-mannheim.de

Abstract. In this paper, we present a novel approach to Word Sense Induction which is based on topic modeling. Key to our methodology is the use of word-topic distributions as a means to estimate sense distributions. We provide these distributions as input to a clustering algorithm in order to automatically distinguish between the senses of semantically ambiguous words. The results of our evaluation experiments indicate that the performance of our approach is comparable to state-of-the-art methods whose sense distinctions are not as easily interpretable.

Keywords: word sense induction, topic models, lexical semantics.

1 Introduction

Computational approaches to the identification of meanings of words in context, a task commonly referred to as Word Sense Disambiguation (WSD) [12], typically rely on a fixed sense inventory such as WordNet [5]. But while Word-Net provides a high-quality semantic lexicon in which fine-grained senses are connected by a rich network of meaningful semantic relations, it is questionable whether or not it provides enough coverage to be successfully leveraged for high-end, real-world applications, e.g., Web search, or whether these need to rely, instead, on sense distinction automatically mined from large text collections [15,4].

An alternative to WSD approaches is offered by methods which aim at automatically discovering senses from word (co-)occurrence in texts, i.e., performing so-called Word Sense Induction (WSI). WSI is viewed as a clustering task where the goal is to assign different occurrences of the same sense of a word to the same cluster and, by converse, to discover different senses of the same word in an unsupervised fashion by assigning their occurrences in text to different clusters. To this end, a variety of clustering methods can be used [12].

All clustering methods, however, crucially depend on the representation of contexts as their input. A standard approach is to view texts simply as vectors of words: a vector space model [17], in turn, can be complemented by dimensionality reduction techniques like, for instance, Latent Semantic Analysis (LSA) [3,18].

* Financed by a Margarete-von-Wrangell scholarship of the European Social Fund (ESF) and the Ministry of Science, Research and the Arts Baden-Württemberg.

I. Gurevych, C. Biemann, and T. Zesch (Eds.): GSCL 2013, LNAI 8105, pp. 97–103, 2013.
© Springer-Verlag Berlin Heidelberg 2013

An alternative method proposed by Brody and Lapata [2], instead, consists of a generative model. In their approach, occurrences of ambiguous words in context are viewed as samples from a multinomial distribution over senses. These, in turn, are generated by sampling a sense from the multinomial distribution and then choosing a word from the sense-context distribution.

All in all, both vector space and generative models achieve competitive performance by exploiting the distributional hypothesis [8], i.e., the assumption that words that occur in similar contexts will have similar meanings. But while distributional methods have been successfully applied to the majority, if not all, of Natural Language Processing (NLP) tasks, they are still difficult to interpret for humans. LSA, for instance, can detect that a word may appear near completely different kind of words, but it does not, and cannot, encode explicitly in its representation that it has multiple senses.[1] In this paper we propose to overcome this problem by exploring the application of a state-of-the-art generative model, namely probabilistic topic models [16] to the task of Word Sense Induction. To this end, we propose to use Topic Models (TMs) as a way to estimate the distribution of word senses in text, and use topic-word distributions as a way to derive a semantic representation of ambiguous words in context that are later clustered to identify their senses. This is related to the approach presented in [10] where each topic is considered to represent a sense, while in this work we use all topics to represent a word's sense.

TMs, in fact, have been successfully used for a variety of NLP tasks, crucially including WSD [1] – thus providing a sound choice for a robust model – and, thanks to their ability to encode different senses of a polysemous word as a distribution over different topics, we expect them to provide a model which is easier to interpret for humans.

2   Method

The intuition behind TMs is that each topic has a predominant theme and ranks words in the vocabulary accordingly. A word's probability in a topic represents its importance with respect to the topic's theme and consequently reflects how dominant the respective theme is for the word. Our assumption in this work is that the meaning of a word consists of a distribution over topics. This assumption is analogue to the distributional hypothesis: words with a similar distribution over topics have similar meanings.

2.1   Representing Word Semantics with Topic Models

A topic model induced from a corpus provides topics $t_1, \ldots, t_i$ and each topic consists of a probability distribution over all words $w_1, \ldots, w_j$ in the vocabulary. Therefore, each word is associated with a probability for a topic, namely

---

[1] Cf., e.g., [7]: "the representation of words as points in an undifferentiated euclidean space makes it difficult for LSA to solve the disambiguation problem".

$p(w_{ji}) = p(w_i|t_j)$.[a] We will represent each word by means of a *topic signature* where the topics define the signature's features, and the word's probabilities in these topics constitute the feature values: $tsig(w_j) = \langle p(w_{j1}), p(w_{j2}), \ldots, p(w_{ji}) \rangle$

The whole document can be represented by aggregating the topic signatures of all the words in the document resulting in a single topic signature that describes the topical focus of the document. This representation of documents can be used to compute euclidean distances between documents which can be input to an established unsupervised clustering algorithm. This fact is utilized to identify word senses.

2.2 Identifying Word Senses

We use the arithmetic mean to aggregate the topic signatures of the words in any given document *doc*.

$$aggr(doc) = \frac{tsig(c_1) + \ldots + tsig(c_m)}{m} \quad \text{where} \quad c_1, \ldots, c_m \in doc \quad (1)$$

We write c when we refer to words in a document (tokens) while w is used for words in the vocabulary (types). Clustering the aggregated document vectors forms groups of documents that are close to each other and thus share similar topic distributions for the contained words. The cluster centroids constitute what we were looking for, a topic based representation of a sense, the *sense blueprints*. Because topics can be interpreted by humans by listing the top words of a topic, the sense blueprints can be interpreted as well. We just look at the top words in each topic along with its amplitude for a single cluster to get an intuition what theme is important in the cluster. This can also be helpful for comparing clusters and identify the topical dimensions where they differ from each other. The complete workflow is depicted in Figure 1.

3 Experiments

For our experiments we use data provided by the WSI task of SemEval 2010 [11]. For each of the 100 ambiguous words (50 nouns and 50 verbs, each having an entry in WordNet) the training data set provides text fragments of 2-3 sentences that were downloaded semi-automatically from the web. Each fragment is a short document that represents one distinct meaning of a target word.

The documents were preprocessed and only the lemmatized nouns were kept to be used as contextual features. We leave the inclusion of verbs for future work, because the average number of senses is higher for verbs than for nouns and thus would introduce more noise [13]. One topic model was built for each

---

[a] Usually the notation includes the information that t_j is sampled from a distribution itself, but as we do not rely on this sampling step in our approach we keep the notation simple.

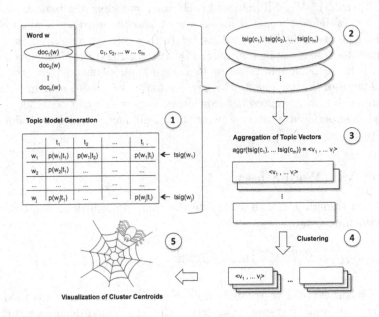

Fig. 1. The complete workflow

ambiguous word on its preprocessed documents using David Blei's TM implementation *ldac*[2]. We set the number of topics to 3, 4, ..., 10 in order to test how the number of topics influences the results. The value for alpha was 0.3.

TMs are used as described in Section 2.1 to represent the documents in the test set as topic signatures. They are clustered using K-means clustering where the number of clusters was determined by the number of WordNet senses of a word.[3]

A spider diagram visualization of two of the cluster results for the word "promotion" can be found in Figure 2. Each dimension corresponds to one topic and is labeled with the respective most probable words. Each cluster centroid is a vector that spans a plane in the diagram, the number of documents per cluster is specified next to the cluster name in the legend. The word "promotion" is dominant in every topic because it appears in every document. With a higher number of topics more thematic details emerge. For example with the number of topics set to 3, there is one topic clearly dealing with advertising (lower right corner) while with 9 topics there are several topics dealing with aspects of advertising.

---

[2] Available at http://www.cs.princeton.edu/~blei/lda-c/index.html

[3] In order to evaluate if the presented approach is worth investigating further we leave the tuning of the number of cluster results for future work. By relying on external resources we avoid introducing more possible variation on the quality of the results by estimating the right number of clusters automatically.

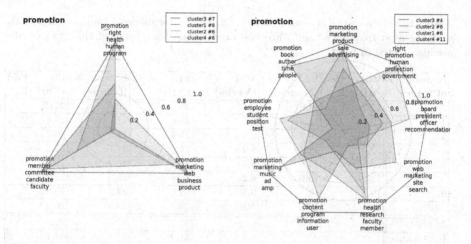

Fig. 2. Clustering result for the word "promotion" with 4 Clusters. The setup on the left side uses 3 topics, the one on the right 9 topics. The cluster labels in the diagrams do not correspond to each other.

4   Results

Following the standard evaluation of the Semeval WSI task, we used paired F-Score and V-measure [14] to evaluate our results. Paired F-Score is the harmonic mean of precision and recall. V-measure is the harmonic mean of the homogeneity and completeness scores of a clustering result. The results of our system are presented in Table 1.

The system's F-Score is not able to outperform the random baseline and does not improve with the number of topics used. Inspecting the detailed results shows that the reason for the significantly lower F-Score lies in low recall values. On average our system has 19% recall in comparison to 99% for MFS while the systems precision is 49% on average in comparison to 39% for MFS. This indicates that the choice of the number of clusters for each word – which was the number of WordNet senses in our experiment – is higher than than the actual number of senses in the gold standard. In fact, the gold standard data generally uses fewer senses than listed by WordNet. Interestingly the F-Score for nouns is similar to the score of verbs, while most other systems reported better results for verbs. The main reason in our opinion is that the training set for nouns was significantly bigger than for verbs, which resulted in more accurate topics. In general working with a corpus of short documents like the Semeval training data makes it harder to identify meaningful topics.

The outcome is much better for V-measure where the results indicate that the system learned useful information because it easily outperforms the random baseline. In comparison to the original results the system would achieve rank four out of 26 systems, with the best 3 systems reporting V-measures of 16.2% (*Hermit*) and 15.7% (both *UoY* and *KSU KDD*). Still, the results are not in the range of recent work like [10].

Table 1. V-measure (VM) and F-Score (FS) results for different topic settings along with the most frequent sense (MFS) and random baseline. The best clustering results are highlighted in **bold**.

Number of topics	VM (%) (All)	VM (%) (Nouns)	VM (%) (Verbs)	FS (%) (All)	FS (%) (Nouns)	FS (%) (Verbs)
3	12.3	14.6	9	27	**26.7**	**27.5**
4	12.6	15.3	8.7	25.6	25.7	25.4
5	12.9	15.2	9.6	25.4	25.9	26.2
6	12.4	14.4	9.4	24.8	24.1	25.7
7	12.5	14.6	9.4	24.8	24.2	25.7
8	13.2	15.9	9.2	25.4	25.2	25.7
9	13	15.5	9.4	24.9	24.3	25.9
10	**14**	**16.7**	**10.1**	25.9	25.5	26.3
MFS	0	0	0	63.5	57	72.2
Random	4.4	4.2	4.6	31.9	30.4	34.1

5   Conclusion and Outlook

In this paper we explored the embedding of information from a generative model in a vector space, in order to create interpretable clustering results. We presented an approach to WSI that uses probabilistic Topic Modeling to create a semantic representation for documents that allows clustering to find word senses. The results do not outperform other approaches to the Semeval 2010 WSI task, but the general idea might be helpful for tasks where interpretable results are desirable like near synonym detection [9] or exploratory data analysis.

There are many directions which we plan to explore in the very near future. Instead of the training set, a big corpus like Wikipedia could be used for creating the topic models. In general we expect the clustering performance to improve when bigger training data sets are available for the topic model creation. In this work, the complete generative model was not incorporated: The infered topic distribution for single documents could be used to add weights to the topic distribution of the words. In order to have a completely unsupervised WSI approach a clustering method that does not need to know the number of clusters beforehand needs to be developed. Additionally hierarchical topic models [6] could be used to find a more fine grained semantic representation.

References

1. Boyd-Graber, J., Blei, D., Zhu, X.: A topic model for word sense disambiguation. In: Proceedings of the Conference on Empirical Methods in Natural Language Processing (EMNLP) and Computational Natural Language Learning (CoNLL), pp. 1024–1033 (2007)
2. Brody, S., Lapata, M.: Bayesian word sense induction. In: Proceedings of the Conference of the European Chapter of the Association for Computational Linguistics (EACL), pp. 103–111 (2009)

3. Deerwester, S., Dumais, S.T., Furnas, G.W., Landauer, T.K., Harshman, R.: Indexing by latent semantic analysis. Journal of the American Society for Information Science 41(6), 391–407 (1990)
4. Di Marco, A., Navigli, R.: Clustering and diversifying web search results with graph-based word sense induction. Computational Linguistics 39(4) (2013)
5. Fellbaum, C.: WordNet: An Electronic Lexical Database (Language, Speech, and Communication). The MIT Press (May 1998)
6. Griffiths, T., Jordan, M., Tenenbaum, J.: Hierarchical topic models and the nested chinese restaurant process. Advances in Neural Information Processing Systems 16, 106–114 (2004)
7. Griffiths, T.L., Steyvers, M., Tenenbaum, J.B.: Topics in semantic representation. Psychological Review 114(2), 211 (2007)
8. Harris, Z.S.: Distributional structure. Word (1954)
9. Hirst, G.: Near-synonymy and the structure of lexical knowledge. In: AAAI Symposium on Representation and Acquisition of Lexical Knowledge: Polysemy, Ambiguity, and Generativity, pp. 51–56 (1995)
10. Lau, J.H., Cook, P., McCarthy, D., Newman, D., Baldwin, T.: Word sense induction for novel sense detection. In: Proceedings of the 13th Conference of the European Chapter of the Association for Computational Linguistics, pp. 591–601. Association for Computational Linguistics, Avignon (2012)
11. Manandhar, S., Klapaftis, I., Dligach, D., Pradhan, S.: Semeval-2010 task 14: Word sense induction & disambiguation. In: Proceedings of the 5th International Workshop on Semantic Evaluation, pp. 63–68. Association for Computational Linguistics, Uppsala (July 2010)
12. Navigli, R.: Word sense disambiguation: A survey. ACM Computing Surveys (CSUR) 41(2), 10 (2009)
13. Ng, H.T.: Getting serious about word sense disambiguation. In: Proceedings of the ACL SIGLEX Workshop on Tagging Text with Lexical Semantics, pp. 1–7 (1997)
14. Rosenberg, A., Hirschberg, J.: V-measure: A conditional entropy-based external cluster evaluation measure. In: Proceedings of the Conference on Empirical Methods in Natural Language Processing (EMNLP), vol. 410, p. 420 (2007)
15. Schuetze, H., Pedersen, J.O.: A cooccurrence-based thesaurus and two applications to information retrieval. Information Processing and Management 33(3), 307–318 (1997)
16. Steyvers, M., Griffiths, T.: Probabilistic topic models. In: Landauer, T., Mcnamara, D., Dennis, S., Kintsch, W. (eds.) Latent Semantic Analysis: A Road to Meaning. Laurence Erlbaum (2007)
17. Turney, P.D., Pantel, P.: From frequency to meaning: Vector space models of semantics. Artificial Intelligence 37(1), 141–188 (2010)
18. Van de Cruys, T., Apidianaki, M., et al.: Latent semantic word sense induction and disambiguation. In: Proceedings of the Annual Meeting of the Association for Computational Linguistics (ACL), pp. 1476–1485 (2011)

Named Entity Recognition in Manipuri:
A Hybrid Approach

Jimmy L and Darvinder Kaur

Department of CSE/IT, Lovely Professional University, Phagwara, Punjab, India
jimmy.laishram@gmail.com,
darvinder14814@lpu.co.in

Abstract. This paper provides a hybrid approach to the Named Entity Recognition (NER) in Manipuri language. The hybrid approach so used is the combination of statistical approach (Conditional Random Field, CRF) and rule-based approach. The rule-based approach helps in defining various unique word features that are used in accurately classifying the Named Entities by the CRF classifier. With small corpus size, this hybrid approach proves to have Recall, Precision and F-score of 92.26%, 94.27% and 93.3% respectively.

Keywords: NER, Machine Learning, Rule Base, Manipuri.

1   Introduction

The Natural Language Processing (NLP) has been a study area since the late 50s, and until now, studies/researches are still in the initial stages for Indian languages because of the many unpredicted language constraints. Named Entity Recognition is one such area of NLP where many approaches have failed to produce results, especially for Indian Languages (ILs).

Named Entity Recognition (NER) is the process of classifying named entities (NE) from a given set of text documents. Such NEs include: person names, location names, organization names, abbreviation, date, currency, numbers, etc.

In the context of ILs, state-of-art accuracy and performance are lacking because of highly complex structure of sentence and word formation. And it is evident that, combination of machine learning approach and rule-based approach only can yield better accuracy [1]. Most of the NERs in ILs are implemented using the machine learning approach [2], [3], [7], [9], [14]. Notable works on the NER system of Manipuri language can be seen in [2-3].

This paper deals with the NER of Manipuri language using the combination of rule-based approach and machine learning approach. The rule-based approach is applied to extract features for Machine learning approach.

1.1   Challenges in Manipuri Language

Manipuri is a Tibeto-Burman language [2], which is scheduled in the Indian Constitution. The following describes the challenges in the language:

I. Gurevych, C. Biemann, and T. Zesch (Eds.): GSCL 2013, LNAI 8105, pp. 104–110, 2013.

- Lack of capitalization for named entities as in English or any other European language.
- Ambiguous meaning of named entities. That is named entities can also found as 'verb' in the dictionary.
- The language has free word order i.e. Subject-Object position can be changed without changing the meaning of the sentence.
- Suffixes define the class of the word.
- Case markers are added to the named entities as suffixes which create difficulties in NER tasks.
- The language is highly inflectional, thus giving challenges in creating set of linguistic rules and statistical features.
- Suffixes can be bundled up one after another to form a complex word.

   Example: পুশিনহঁনজঁরঁমগঁদঁবঁনিদঁকো (pushinhƏnjƏrƏmgƏdƏbƏnidƏko) [3]
   (I wish I have made him/her brought inside)
   পু-শিন-হ্ন-জঁ-রম-গা-দা-বা-নি-দা-কো
   (Pu-shin-hƏn-jƏ-rƏm-gƏ-dƏ-bƏ-ni-dƏ-ko)

The word contains 10 (ten) suffixes which are added to the verbal root পু (Pu). The above example proves how agglutinative the Manipuri language is.

- Named entities can also be found as affixes in the list which creates complexity in stemming.
- The Manipuri language has very limited resources such as annotated corpus, POS tagger, stemmer, etc.

2   Conditional Random Field

The Conditional Random Field is a machine learning approach (supervised learning technique) which calculates a conditional probability $P(Y|x)$ over label sequence given a particular observation x [3].

Let X be the random variable over a data sequence to be labelled and Y be the random variable over corresponding label sequence. In our case, X ranges over natural language sentence and Y is the set of possible NE tags.

The conditional probability of state sequence $Y= \{y_1, y_2, y_2... y_n\}$ given the observation $X= \{x_1, x_2, x_3... x_n)$ is calculated as

$$P(Y|X) = \exp\left(\sum_j \lambda jtj(yi - 1, yi, x, i) + \sum_k uk, sk(yi, x, i)\right)$$

where $\lambda jtj(yi - 1, yi, x, i)$ is the transition feature function of the entire observation sequence and the labels at position i-1 and i. $sk(yi, x, i)$ is the state feature function of the label at position i and the observation sequence. λj and uk are the parameters to be estimated from the training data.

Now, for each observation, we can construct a set of real-valued feature B(x, i) such that

$$B(x, i) = \begin{cases} 1, \text{if x at position i is the word "lastname"} \\ 0, \text{otherwise} \end{cases}$$

$B(x, i)$ expresses some characteristics of the empirical distribution and should also hold for the model distribution.

The feature function can be defined as:

$$tj\,(yi - 1, yi, x, i) = \begin{cases} B(x,i), \text{if } yi - 1 = B - PER \text{ and } yi = E - PER \\ 0, \text{otherwise} \end{cases}$$

Rewriting the probability of label sequence Y over the observation sequence X

$$P\,(Y|X, \lambda) = 1/Z(x) \exp\left(\sum \lambda_j F_j(Y, X)\right) \tag{1}$$

Where $F_j\,(Y, X) = \sum_{j=1}^{n} fj\,(yi - 1, yi, x, i)$ and $Z(x)$ = Normalization Factor which makes the probability of the entire state sequence sum to 1.

Now assuming that the training data {Xk, Yk} are independently and identically distributed, the product of (1) over the training sequence, as a function of parameter λ, is the likelihood denoted by P (Yk | Xk, λ).

The log-likelihood is given by

$$\sum_k (\log\,(P\,(Yk|Xk, \lambda)) \tag{2}$$

Equation 2 is a concave function, guaranteeing convergence to the global maximum. The maximum likelihood training chooses parameter values in such a manner that (2) is maximized.

3   Named Entity Recognition in Manipuri

The corpus is manually tagged with the corresponding NE tags, which contains nearly 40000 word-forms, in which about 10000 words are Named Entities. Table 1 describes the used NE Tagset.

Table 1. Named Entity Tag-sets

NE Tag	Meaning	Example
LOC	Single word location name	ইম্ফাল(Imphal)
PER	Single word person name	রাজু (Raju)
ORG	Single word organization name	অমাদা (AMADA)
B-LOC I-LOC E-LOC	Beginning, Internal and End of a Multi-word location name	মনিঙ (Maning) লাংজম (Longjam) লৈকাই (Leikai)
B-PER I-PER E-PER	Beginning, Internal and End of a Multi-word person name	হুইদ্রোম(Huidrom) ইন্দুমালা (Indumala) দেবী (Devi)
B-ORG I-ORG E-ORG	Beginning, Internal and End of a Multi-word organization name	ইমা (Ima) য়াইফরেল (Yaipharel) লুপ (Lup)

Table 1. (*continued*)

B-CUR I-CUR E-CUR	Beginning, Internal and End of a Multi-word Currency	লুপা (Lupa) ক্রোর (Crore) অমা (Ama)
B-DAT I-DAT E-DAT	Beginning, Internal and End of a Multi-word Date	2006 ডিসেম্বর (December) 25
B-TIM E-TIM	Beginning and End of a Multi-word time	পুং (Pung) ১০(10)

For research, C++ based CRF++ 0.57 package has been used for labelling and segmenting of data, which is a readily available open source and Java for stemming and feature creation.

Figure 1 shows the working process of this NER system.

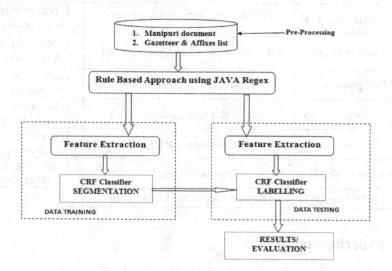

Fig. 1. Hybrid system of NER Manipuri

3.1 Gazetteer Lists and Stemming Affixes

A gazetteer list is maintained for last-names, common location name, designation, etc. Such lists are finite and contain only those words which are unchangeable. The gazetteer list helps in creating a unique identifiable feature for each multiword named entity. The features having binary notation in table 2 are created using the gazetteer list.

The words have been stemmed by removing the suffixes and the prefixes. Suffixes and prefixes lists are maintained and by using JAVA Regex the affixes are stripped. A training file is generated using the gazetteer list and stemming process. The training file consists of tokens described in table 2.

3.2   CRF Features

The followings are the features used for the machine learning:

Table 2. CRF Features

FEATURES	MEANING	BINARY
Word-[prefix/suffix]	Word after removing Prefixes/ Suffixes	NA
Prefix	Represents the removed prefix	NA
Suffix	Represents the removed suffixes	NA
Sur-name indicator	Indicates the word is a Sur-name	1, occurs in list 0, otherwise
First-name indicator	Indicates the word is a First-name	1, occurs in list 0, otherwise
Last-name indicator	Indicates the word is a Last-name	1, occurs in list 0, otherwise
Location indicator	Indicates the word is a location name	1, occurs in list 0, otherwise
Designation	Represents the designation	1, occurs in list 0, otherwise
Date	Represents date	1, occurs in list 0, otherwise
Currency	Represents currency	1, occurs in list 0, otherwise
Number in words	Represents number	1, occurs in list 0, otherwise
POS-Tag	Part-of-speech tag of the word	NA

4   Experimental Results

4.1   Selection of Best Features

Kishorjit Nongmeikappam et al, "*CRF based Named Entity Recognition NER in Manipuri: A Highly Agglutinative language*", IEEE 0.1109/NCETACS.2011, reports that the best range for considering surrounding words is -2 to +2. That is $W[i-2]$, $W[i-1]$, $W[i]$, $W[i+1]$, $W[i+2]$, where W= word. The research gave an overall F-score of 83.33%. So this surrounding word window is considered in this research.

The features (F) used for the CRF training in this research are:

F= {W[i], SW[i], W[i-2], W[i-1], W[i+1], W[i+2], S[i], P[i], SNb[i], FNb[i], LNb[i], LCNb[i], Db[i], Cb[i], Nb[i], POS[i], POS[i-2], POS[i-1], POS[i+1], POS[i+2] }

Where

Table 3. Notations and meanings

Notation	Meaning
W[i, j]	Current word spanning from ith left position to jth right position
SW[i, j]	Stem word spanning from ith left position to jth right position
S[i]	Suffix at position i
P[i]	Prefix at Position i
SNb[i]	Surname binary notation at position i
FNb[i]	First-name binary notation at position i
LNb[i]	Last-name binary notation at position i
LCNb[i]	Location name binary notation at position i
Db[i]	Date binary notation at position i
Cb[i]	Currency binary notation at position i
Nb[i]	Number binary notation at position i
POS[i, j]	POS spanning from ith left position to jth right position

4.2 Experimental Values

The corpus contains about 40000 word-forms. 30000 word-forms have been used as training set. The system has been tested with 10247 words having unique 1024 Named Entities. The result so obtained is shown below.

Table 4. Results

MODEL	RECALL	PRECISION	F-SCORE
CRF+ Rule Based (Gazetteer list)	92.26	94.27	93.3

5 Conclusion

In this paper, a Named Entity Recognition system has been created using a hybrid approach which is a combination of statistical approach and rule-based approach. The CRF is used because of the reason that CRF does not suffer from **label bias problems** due to unseen data in training. The result so obtained shows that hybrid approach yields state of the art accuracy with small corpus size. The feature selection is done by the rule-based approach, which can be further enhanced by using genetic algorithms. Unsupervised statistical approach for NER could be on the future road.

References

1. Srivastava, S., Sanglikar, M., Kothari, D.C.: Named Entity Recognition System for Hindi language: A Hybrid Approach. In: The Proceedings of the IJCL-2011, pp. 10–23 (2011)
2. Singh, T.D., Nongmeikappam, K., Ekbal, A., Bandyopadhyay, S.: Named Entity Recognition for Manipuri using SVM. In: The Proceedings of 23rd Pacific Asia Conference on Language, Information and Computation, pp. 811–818 (2009)

3. Nongmeikappam, K., Singh, L.N., Shangkhunem, T., Salam, B., Chanu, N.M., Bandhyopadhyay, S.: CRF based Named Entity Recognition in Manipuri: A Highly Agglutinative Language, pp. 92–97. IEEE 0.1109/NCETACS (2011)
4. Saha, S.K., Sarkar, S., Mitra, P.: Gazetteer Preparation of Named Entity Recognition in Indian Language. In: The Proceedings of the 6th Workshop on Asian Language Resource, pp. 9–16 (2008)
5. Kaur, D., Gupta, V.: A Survey of Named Entity Recognition in English and other Indian Language. In: The Proceedings of the IJCSI, pp. 239–245 (2010)
6. Sashidhar, B., Yohan, P.M., Babu, A.V., Govardhan, A.: A Survey on Named Entity Recognition in Indian Language with particular reference to Telegu. In: The Proceedings of the IJCSI, pp. 438–443 (2011)
7. Singh, T.D., Bandyopadhyay, S.: Web Based Manipuri Corpus for Mutltiword NER and Reduplicated MWEs Identification using SVM. In: The Proceedings of the 1st Workshop on South and Southeast Asian Natural Language Processing, pp. 35–42 (2010)
8. Biswas, S., Mishra, S.P., Acharya, S., Mohanty, S.: Hybrid Oriya Named Entity Recognition system: Harnessing the Power of Rule. The Proceedings of the International Journal of Artificial Intelligence and Expert System, 1–6 (2010)
9. Ekbal, A., Bandyopadhyay, S.: Named Entity Recognition using Appropriate Unlabeled Data, Post-processing and Voting. In: The Proceedings of Informatica, pp. 2–3 (2010)
10. Petasis, G., Petridis, S., Paliouras, G., Karkaletsis, V., Perantonis, S.J., Spyropoulos, C.D.: Symbolic and Neural Learning for Named Entity Recognition. In: The Proceedings of 5th International Conference MTSR (2011)
11. Klien, D., Smarr, J., Nguyen, H., Manning, C.D.: Named Entity Recognition with Character-Level Models. In: The Proceedings of the 7th Conference on Natural Language Learning, pp. 180–183 (2003)
12. Chieu, H.L., Ng, H.T.: Named Entity Recognition: A Maximum Entropy Approach using Global Information. In: The Proceedings of the 19th International Conference on Computational Linguistic (2002)
13. Hasan, K.S., Rahman, A., Ng, V.: Learning-Based Named Entity Recognition for Morphologically-Rich Resource-Scarce Languages. In: The Proceedings of 12th Conference of the European Chapter of the Association for Computational Linguistics (EACL), Athens, Greece, pp. 354–362 (2009)
14. Ekbal, A., Haque, R., Bandyopadhyay, S.: Named Entity Recognition in Bengali: A Conditional Random Field Approach. In: The Proceedings of International Joint Conference on Natural Language Processing, pp. 589–594 (2008)
15. Ekbal, A., Bandyopadhyay, S.: Named Entity Recognition in Bengali: A Multi-Engine Approach. In: The Proceedings of Northern European Journal of Language Technology, pp. 26–48 (2009)

A Study of Chinese Word Segmentation Based on the Characteristics of Chinese

Aaron Li-Feng Han, Derek F. Wong, Lidia S. Chao,
Liangye He, Ling Zhu, and Shuo Li

University of Macau, Department of Computer and Information Science
Av. Padre Toms Pereira Taipa, Macau, China
{hanlifengaaron,wutianshui0515,imecholing,leevis1987}@gmail.com,
{derekfw,lidiasc}@umac.mo

Abstract. This paper introduces the research on Chinese word segmentation (CWS). The word segmentation of Chinese expressions is difficult due to the fact that there is no word boundary in Chinese expressions and that there are some kinds of ambiguities that could result in different segmentations. To distinguish itself from the conventional research that usually emphasizes more on the algorithms employed and the workflow designed with less contribution to the discussion of the fundamental problems of CWS, this paper firstly makes effort on the analysis of the characteristics of Chinese and several categories of ambiguities in Chinese to explore potential solutions. The selected conditional random field models are trained with a quasi-Newton algorithm to perform the sequence labeling. To consider as much of the contextual information as possible, an augmented and optimized set of features is developed. The experiments show promising evaluation scores as compared to some related works.

Keywords: Natural language processing, Chinese word segmentation, Characteristics of Chinese, Optimized features.

1 Introduction

With the rapid development of natural language processing and knowledge management, word segmentation becomes more and more important due to its crucial impacts on text mining, information extraction and word alignment, etc. The Chinese word segmentation (CWS) faces more challenges because of a lack of clear word boundary in Chinese texts and the many kinds of ambiguities in Chinese. The algorithms explored on the Chinese language processing (CLP) include the Maximum Matching Method (MMM) [1], the Stochastic Finite-State model [2], the Hidden Markov Model [3], the Maximum Entropy method [4] and the Conditional Random Fields [5] [6], etc. The workflow includes self-supervised models [7], unsupervised models [8] [9], and a combination of the supervised and unsupervised methods [10]. Generally speaking, the supervised methods gain higher accuracy scores than the unsupervised ones. The Hidden Markov Model

I. Gurevych, C. Biemann, and T. Zesch (Eds.): GSCL 2013, LNAI 8105, pp. 111–118, 2013.

(HMM) assumes a strong independence between variables forming an obstacle for the consideration of contextual information. The Maximum Entropy (ME) model has the label bias problem seeking the local optimization rather than global. Conditional random fields (CRF) overcome the above two problems but face new challenges, such as the selection of the optimized features for some concrete issues. Furthermore, the conventional research tends to emphasize more on the algorithms utilized or the workflow designed while the exploration of the essential issues under CWS is less mentioned.

2 Characteristics of Chinese

2.1 Structural Ambiguity

The structural ambiguity phenomenon usually exists between the adjacent words in a Chinese sentence. This means one Chinese character can be combined with the antecedent characters or subsequent characters, and both combinations result in reasonable Chinese words. The structural ambiguity leads to two problems.

a). It results in two differently segmented sentences, and each of the sentences has a correct language structure and sensible meaning. For instance, the Chinese sentence "他的船只靠在維多利亞港" has a structural ambiguity between the adjacent words "船只靠在". The Chinese character "船" and the combined characters "船只" have the same meaning, "ship" as a noun. The Chinese character "只" also has the meaning of "only" as an adverb. So the adjacent characters "船只靠在" could be segmented as "船只/靠在" (a ship moors in) or "船/只/靠在" (ships only moor in). Thus, the original Chinese sentence has two different structures and the corresponding meanings "他的/船只/靠在/維多利亞港" (His ship is mooring in Victoria Harbour) and "他的/船/只/靠在/維多利亞港" (His ship only moors in Victoria Harbour).

b). It results in two differently segmented sentences with one sentence having a correct structure while the other one being wrongly structured and not forming a normal Chinese sentence. For instance, the Chinese sentence "水快速凍成了冰" has a structural ambiguity between the characters "快速凍". The combined characters "快速" mean "fast" as an adjective or an adverb, while "速凍" is also a Chinese word usually as an adjective to specify some kind of food, e.g. "速凍/食品" (fast-frozen food). So the original Chinese sentence may be automatically segmented as "水/快速/凍/成了/冰" (the water is quickly frozen into ice.) or "水/快/速凍/成了/冰" (water / fast /fast frozen / into / ice). The second segmentation result does not have a normal Chinese structure.

2.2 Abbreviation and Ellipsis

The two phenomena of abbreviation and ellipsis usually lead to ambiguity in Chinese sentences. Firstly, let's see the abbreviation phenomenon in Chinese human name entities. People sometimes use the Chinese family name to represent one person with his or her given name omitted in the expression. For instance, the Chinese sentence " 許又從街坊口中得知" can be understood as "許又/從/街坊/口中/得知" (XuYou heard from the neighbors) with the word "許又" (XuYou) as a Chinese full name because "許" (Xu) is a commonly used Chinese family name which is to be followed by one or two characters as the given name. However, the sentence can also be structured as " 許/又/從/街坊/口中/得知" (Xu once more heard from the neighbors) with the surname "許" (Xu) representing a person and the character " 又" (once more) as an adverb to describe the verb "得知" (heard).

Secondly, let's see the ellipsis phenomenon in place name entities (including the translated foreign place names). People usually use the first Chinese character (the beginning character) of a place name entity to stand for the place especially when the string length of the entity is large (four or more characters). For example, the Chinese sentence "敵人襲擊巴西北部" could be understood as "敵人/襲擊/巴/西北部" (The enemy attacks the northwestern part of Pakistan) with the character "巴" (Ba) as the representation of "巴基斯坦" (Pakistan) which is very common in Chinese international news reports, and the word "西北部" (northwestern part) is a noun of locality. However, the sentence can also be understood as "敵人/襲擊/巴西/北部" (the enemy attacks the northern Brazil) because the two characters "巴西" (Brazil) combine to mean the Chinese name of Brazil and the following characters " 北部" (northern part) combine to function as a noun of locality. Each of these two kinds of segmentations results in a well structured Chinese sentence.

3   CRF Model

The CRFs are both conditional probabilistic models and statistics-based undirected graph models that can be used to seek the globally optimized optimization results when dealing with the sequence labeling problems [11]. Assume X is a variable representing input sequence, and Y is another variable representing the corresponding labels to be attached to X. The two variables interact as conditional probability $P(Y|X)$ mathematically. Then comes the definition of CRF: Let a graph model $G = (V, E)$ comprise a set V of vertices or nodes together with a set E of edges or lines and $Y = \{Y_v | v \in V\}$, such that Y is indexed by the vertices of graph G, then (X, Y) shapes a CRF model. This set meets the following form:

$$P_\theta(Y|X) \propto \exp\left(\sum_{e \in E, k} \lambda_k f_k(e, Y|_e, X) + \sum_{v \in V, k} \mu_k g_k(v, Y|_v, X)\right) \quad (1)$$

where X and Y represent the data sequence and label sequence respectively; f_k and g_k are the features to be defined; λ_k and μ_k are the parameters trained from the data sets; the bar "|" is the mathematical symbol to express that the right part is the precondition of the left. The training methods for the parameters include the Conjugate-gradient algorithm [12], the Iterative Scaling Algorithms

[11], the Voted Perceptron Training [13] etc. In this paper we select a quasi-newton algorithm [14] and some online tools[1] to train the model.

4   Features Designed

As discussed in the previous sections of this paper about the characteristics of Chinese, CWS is highly reliant on the contextual information to deal with the ambiguity phenomena and yield the correct segmentations of sentences. So we will employ a large set of features to consider as much of the contextual information as possible. Furthermore, the name entities play an important role in sentence segmentations [15]. For example, the Chinese sentence "新疆維吾爾 自治區/分外/妖嬈" (the Xinjiang Uygur Autonomous Region is extraordinarily enchanting) is probable wrongly segmented as " 新疆/維吾爾/自治/區分/外/妖 嬈" (Xinjiang / Uygur / autonomy / distinguish / out / enchanting) due to that the location name entity "新疆維吾爾自治區" (Xinjiang Uygur Autonomous Region) is a long string (8 characters) and its last character " 區" (Qu) can be combined with the following character " 分" (Fen) to form a commonly used Chinese word " 區分" (distinguish).

There are several characteristics of Chinese name entities. First, some of the name entities contain more than five or six characters that are much longer than the common names, e.g. the location name "古爾班通古特沙漠" (Gu Er Ban Tong Gu Te Desert) and the organization name "中華人民共和國" (People's Republic of China) contain eight and seven Chinese characters respectively. Second, one name entity usually can be understood as composed by two inner parts i.e. its proprietary name (in the beginning of the name entity) and the commonly used categorical name which serves as the suffix, and the suffixes usually contain one or two characters thus are shorter than the proprietary names. Thus, in the designing of the features, we put more consideration into the antecedent characters than the subsequent characters for each token studied to avoid unnecessary consideration of the subsequent characters which may generate noises in the model. The final optimized feature set used in the experiment is shown in Table 1 which includes unigram (U), bigram (B) and trigram (T) features.

5   Experiments

5.1   Corpora

The corpora used in the experiments are CityU (City University of Hong Kong corpus) and CTB (Pennsylvania Chinese Tree Library corpus) which we select from the SIGHAN (a special interest group in Association of Computational Linguistics) Chinese Language Processing Bakeoff-4 [16] for the testing of traditional Chinese and simplified Chinese respectively. CityU and CTB contain 36,228 and 23,444 Chinese sentences respectively in the training corpora, while the corresponding test corpora contain 8,094 and 2,772 sentences.

---

[1] http://crfpp.googlecode.com/svn/trunk/doc/index.html

Table 1. Designed feature sets

Features	Meaning
$U_n, n \in (-4, 1)$	Unigram, from antecedent 4th to subsequent 1st character
$B_{n,n+1}, n \in (-2, 0)$	Bigram, three pairs of characters from the antecedent 2nd to the subsequent 1st
$B_{-1,1}$	Jump Bigram, the antecedent 1st and subsequent 1st character
$T_{n,n+1,n+2}, n \in (-2, -1)$	Trigram, the characters from the antecedent 2nd to the following 1st

Table 2. Information of the test corpora

	Number of Words		
Type	Total	IV	OOV
CityU	235,631	216,249	19,382
CTB	80,700	76,200	4,480

The detailed information of the testing corpora is shown in Table 2. IV (in vocabulary) means the number of words existing in the training corpus while OOV means the number of words never appears in the training corpus.

5.2 Experiment Results

The evaluation scores of experiments are shown in Table 3 with three indicators. The evaluation scores on precision and recall are similar without big differences, leading to the total F-scores 92.68% and 94.05% respectively on CityU and CTB. On the other hand, the evaluation scores also demonstrate that the OOV words are the main challenges in the CWS research (64.20% and 65.62% respectively of the F-score on OOV words).

Table 3. Evaluation scores of results

	Total			IV			OOV		
	Recall	Precision	F	Recall	Precision	F	Recall	Precision	F
CityU	0.9271	0.9265	0.9268	0.9490	0.9593	0.9541	0.6828	0.6057	0.6420
CTB	0.9387	0.9423	0.9405	0.9515	0.9666	0.9590	0.7214	0.6018	0.6562

5.3 Comparison with Related Works

Lu [17] conducted the CWS research using synthetic word analysis with tree-based structure information and annotation of morphologically derived words from training dictionary in **closed track** (only using the information found in the provided training data). Zhang and Sumita [18] created a new Chinese morphological analyzer Achilles by integrating rule-based, dictionary-based, and statistical machine learning methods and CRF. Keong [19] performed CWS using a Maximum Entropy based Model. Wu [20] employed CRF as the primary model and adopts a transformation based learning technique for the post-processing in the **open track** (any external data could be used in addition to the provided training data). Qin [21] proposed a segmenter using forward-backward maximum matching algorithm as the basic model with the external knowledge of the news wire of Chinese People's Daily in the year of 2000 to achieve the disambiguation.

Table 4. Comparison with related works

	Track	IV F-score		OOV F-score		Total F-score	
		CityU	CTB	CityU	CTB	CityU	CTB
[17]	Closed	0.9483	0.9556	0.6093	0.6286	0.9183	0.9354
[19]	Closed	0.9386	0.9290	0.5234	0.5128	0.9083	0.9077
[18]	Closed	0.9101	0.8939	0.6072	0.6273	0.8850	0.8780
[20]	**Open**	0.9401	**0.9753**	0.6090	**0.8839**	0.9098	**0.9702**
[21]	Open	N/A	0.9398	N/A	0.6581	N/A	0.9256
Ours	**Closed**	**0.9541**	**0.9590**	**0.6420**	**0.6562**	**0.9268**	**0.9405**

The comparison with related works shows that this paper yields promising results without using external resources. The closed-track total F-score on CityU corpus (0.9268) even outperforms [20] what is tested on open track using external resources (0.9098).

6   Conclusion and Future Works

This paper focuses on the research work of CWS that is a difficult problem in NLP literature due to the fact that there is no clear boundary in Chinese sentences. Firstly, the authors introduce several kinds of ambiguities inherent with Chinese that underlie the thorniness of CWS to explore the potential solutions. Then the CRF model is employed to conduct the sequence labeling. In the selection of features, in consideration of the excessive length of certain name entities and for the disambiguation of some sentences, an augmented and optimized feature set is designed. Without using any external resources, the experiments yield promising evaluation scores as compared to some related works including both those using the closed track and those using the open track.

Acknowledgments. The authors wish to thank the anonymous reviewers for many helpful comments and Mr. Yunsheng Xie who is a Master Candidate of Arts in English Studies from FSH at the University of Macau for his kind help in improving the language quality.

References

1. Pak-kwong, W., Chorkin, C.: Chinese word segmentation based on maximum matching and word binding force. In: Proceedings of the 16th Conference on Computational Linguistics, COLING 1996, vol. 1, pp. 200-203. Association for Computational Linguistics, Stroudsburg (1996)
2. Richard, S., Willian, G., Chilin, S., Nancy, C.: A stochastic finite-state word-segmentation algorithm for Chinese. Computational Linguistics 22(3), 377–404 (1996)
3. Hua-Ping, Z., Qun, L., Xue-Qi, C., Hao, Z., Hong-Kui, Y.: Chinese lexical analysis using hierarchical hidden Markov model. In: Proceedings of the Second SIGHAN Workshop on Chinese Language Processing, SIGHAN 2003, vol. 17, pp. 63–70. Association for Computational Linguistics, Stroudsburg (2003)
4. Jin, L.K., Hwee, N.T., Wenyuan, G.: A maximum entropy approach to Chinese word segmentation. In: Proceedings of the Fourth SIGHAN Workshop on Chinese Language Pro-cessing, vol. 164 (2005)
5. Fuchun, P., Fangfang, F., An-drew, M.: Chinese segmentation and new word detection using conditional random fields. In: Proceedings of the 20th International Conference on Computational Linguistics (COLING 2004), vol. Article 562. Association for Computational Linguistics, Stroudsburg (2004)
6. Ting-hao, Y., Tian-Jian, J., Chan-hung, K., Richard, T.: T-h., Wen-lian, H.: Unsupervised overlapping feature selection for conditional random fields learning in Chinese word segmentation. In: Proceedings of the 23rd Conference on Computational Linguistics and Speech Processing, ROCLING 2011, pp. 109–122. Association for Computational Linguistics, Stroudsburg (2011)
7. Fuchun, P., Xiangji, H., Dale, S., Nick, C.-C., Stephen, R.: Using self-supervised word segmentation in Chinese in-formation retrieval. In: Proceedings of the 25th An-nual International ACM SIGIR Conference on Research and Development in Information Retrieval (SIGIR 2002), pp. 349–350. ACM, New York (2002)
8. Hanshi, W., Jian, Z., Shiping, T., Xiaozhong, F.: A new unsupervised approach to word segmentation. Computational Linguistics 37(3), 421–454 (2011)
9. Yan, S., Chunyu, K., Ruifeng, X., Hai, Z.: How unsupervised learning affects character tagging based Chinese Word Segmentation: A quantitative investigation. International Conference on Machine Learning and Cybernetics 6, 3481–3486 (2009)
10. Hai, Z., Chunyu, K.: Integrating unsupervised and supervised word segmentation: The role of goodness measures. Information Sciences 181(1), 163–183 (2011)
11. John, L., Andrew, M., Ferando, P.C.N.: Conditional random fields: Probabilistic models for segmenting and labeling sequence data. In: Proceeding of 18th International Conference on Machine Learning, pp. 282–289 (2001)
12. Shewchuk, J.R.: An introduction to the conjugate gradient method without the agonizing pain. Technical Report CMUCS-TR-94-125. Carnegie Mellon University (1994)
13. Michael, C., Nigel, D., Florham, P.: New ranking algorithms for parsing and tagging: kernels over discrete structures, and the voted perceptron. In: Proceedings of the 40th Annual Meeting on Association for Computational Linguistics (ACL 2002), pp. 263–270. Association for Computational Linguistics, Stroudsburg (2002)
14. The Numerical Algorithms Group: E04 - Min-imizing or Maximizing a Function, NAG Library Manual, Mark 23 (2012) (retrieved)
15. Peng, L., Liu, Z., Zhang, L.: A Recognition Approach Study on Chinese Field Term Based Mutual Information /Conditional Random Fields. In: 2012 International Workshop on Information and Electronics Engineering, pp. 1952–1956 (2012)

16. Guangjin, J., Xiao, C.: The Fourth International Chinese Language Processing Bakeoff: Chinese Word Segmentation, Name Entity Recognition and Chinese POS Tagging. In: Proceedings of the Sixth SIGHAN Workshop on Chinese Language Processing, Hyderabad, India, pp. 83–95 (2008)
17. Asahara, L.J.M., Matsumoto, Y.: Analyzing Chinese Synthetic Words with Tree-based Information and a Survey on Chinese Morphologically Derived Words. In: Proceedings of the Sixth SIGHAN Workshop on Chinese Language Processing, Hyderabad, India, pp. 53–60 (2008)
18. Zhang, R., Sumita, E.: Achilles: NiCT/ATR Chinese Morphological Analyzer for the Fourth Sighan Bakeoff. In: Proceedings of the Sixth SIGHAN Workshop on Chinese Language Processing, Hyderabad, India, pp. 178–182 (2008)
19. Leong, K.S., Wong, F., Li., Y., Dong, M.: Chinese Tagging Based on Maximum Entropy Model. In: Proceedings of the Sixth SIGHAN Workshop on Chinese Language Processing, Hyderabad, India, pp. 138–142 (2008)
20. Wu, X., Lin, X., Wang, X., Wu, C., Zhang, Y., Yu, D.: An Im-proved CRF based Chinese Language Processing System for SIGHAN Bakeoff 2007. In: Proceedings of the Sixth SIGHAN Workshop on Chinese Language Processing, Hyderabad, India, pp. 155–160 (2008)
21. Qin, Y., Yuan, C., Sun, J., Wang, X.: BUPT Systems in the SIGHAN Bakeoff 2007. In: Proceedings of the Sixth SIGHAN Workshop on Chinese Language Processing, Hyderabad, India, pp. 94–97 (2008)

Phrase Tagset Mapping for French and English Treebanks and Its Application in Machine Translation Evaluation

Aaron Li-Feng Han, Derek F. Wong, Lidia S. Chao,
Liangye He, Shuo Li, and Ling Zhu

University of Macau, Department of Computer and Information Science
Av. Padre Toms Pereira Taipa, Macau, China
{hanlifengaaron,wutianshui0515,leevis1987,imecholing}@gmail.com,
{derekfw,lidiasc}@umac.mo

Abstract. Many treebanks have been developed in recent years for different languages. But these treebanks usually employ different syntactic tag sets. This forms an obstacle for other researchers to take full advantages of them, especially when they undertake the multilingual research. To address this problem and to facilitate future research in unsupervised induction of syntactic structures, some researchers have developed a universal POS tag set. However, the disaccord problem of the phrase tag sets remains unsolved. Trying to bridge the phrase level tag sets of multilingual treebanks, this paper designs a phrase mapping between the French Treebank and the English Penn Treebank. Furthermore, one of the potential applications of this mapping work is explored in the machine translation evaluation task. This novel evaluation model developed without using reference translations yields promising results as compared to the state-of-the-art evaluation metrics.

Keywords: Natural language processing, Phrase tagset mapping, Multilingual treebanks, Machine translation evaluation.

1 Introduction

To promote the development of syntactic analysis technology, treebanks for different languages were developed during the past years, such as the English Penn Treebank [1][2], the German Negra Treebank [3], the French Treebank [4], the Chinese Sinica Treebank [5], etc. These treebanks use their own syntactic tagsets, with the number of tags ranging from tens (e.g. the English Penn Treebank) to hundreds (e.g. the Chinese Sinica Treebank). These differences make it inconvenient for other researchers to take full advantages of the treebanks, especially when they undertake the multilingual or cross-lingual research. To bridge the gap between these treebanks and to facilitate future research, such as the unsupervised induction of syntactic structures, Petrov et al. [6] developed a universal part-of-speech (POS) tagset and the POS tags mapping between multilingual

I. Gurevych, C. Biemann, and T. Zesch (Eds.): GSCL 2013, LNAI 8105, pp. 119–131, 2013.

treebanks. However, the disaccord problem in the phrase level tags remains unsolved. Trying to bridge the phrase level tags of multilingual treebanks, this paper
designs a mapping work of phrase tags between French Treebank and English
Peen Treeebank (I and II). Furthermore, this mapping work has been applied
into the machine translation evaluation task (one of the potential applications)
to show its advantages.

2   Designed Phrase Mapping

The designed phrase tags mapping between English and French treebanks is
shown in Table 1. There are 9 universal phrasal categories covering the 14 phrase
tags of Peen Treebank I, 26 phrase tags in Penn Treebank II and 14 phrase tags
in the French Treebank. The brief analysis of the phrase tags mapping will be
introduced as below.

Table 1. Phrase tagset mapping for French and English Treebanks

Universal tags	English Penn Treebank I [1]	English Penn Treebank II [7]	French Tree-bank [4]	Tags Meaning
NP	NP, WHNP	NP, NAC, NX, WHNP, QP	NP	Noun phrase
VP	VP	VP	VN, VP, VPpart, VPinf	Verbal phrase
AJP	ADJP	ADJP, WHAD-JP	AP	Adjectival phrase
AVP	ADVP, WHAD-VP	ADVP, WHAVP, PRT	AdP	Adverbial phrase
PP	PP, WHPP	PP, WHPP	PP	Prepositional phrase
S	S, SBAR, SBAR-Q, SINV, SQ	S, SBAR, SBAR-Q, SINV, SQ, PRN, FRAG, RRC	SENT, Ssub, Sin-t, Srel, S	(sub-)Sentence
CONJP		CONJP		Conjunction phrase
COP		UCP	COORD	Coordinated phrase
X	X	X, INTJ, LST		Others

The universal phrase tag NP (noun phrases) covers: the French tag NP (noun
phrases); the English tags NP (noun phrases), NAC (not a constituent, used to
show the scope of certain prenominal modifiers within an NP), NX (used within
certain complex NPs to mark the head of the NP), WHNP (wh-noun phrase,
containing some wh-word, e.g. who, none of which) and QP (quantifier phrase).
There are no corresponding French tags that could be aligned to English tag
QP. On the other hand the corresponding quantifier phrase is usually marked

as NP in French, so the English tag QP is classified into the NP category in the mapping, for instance:

English: (QP (CD 200) (NNS millions))
French: (NP (D 200) (N millions))

The universal phrase tag VP (verb phrase) covers: the French tags VN (verbal nucleus), VP (infinitives and nonfinite clauses, including VPpart and VPinf); the English tag VP (verb phrase). In English, the infinitive phrase (to + verb) is labeled as VP, so the French VP (infinitives and nonfinite clauses) is classified into the same category, for instance (English infinitive):

(VP (TO to)
    (VP (VB come)))

The universal phrase tag AJP (adjective phrase) covers: the French tag AP (adjectival phrases); the English tags ADJP (adjective phrase), WHADJP (wh-adjective phrase, e.g. how hot).

The universal phrase tag AVP (adverbial phrase) contains: the French tag AdP (adverbial phrases); the English tags ADVP (adverb phrase), WHAVP (wh-adverb phrase) and PRT (particle, a category for words that should be tagged RP). The English phrase tag PRT is classified into the adverbial phrase tag because it is usually used to label the word behind an English verb and there is no similar phrase tag in the French Treebank to align it, for instance (English PRT):

(VP (VBG rolling)
    (PRT (RP up)))

The universal phrase tag PP (prepositional phrase) covers: the French tag PP (prepositional phrases); the English tags PP (prepositional phrase) and WHPP (wh-prepositional phrase).

The universal phrase tag S (sentence and sub-sentence) covers: the French tags SENT (sentences), S (finite clauses, including Ssub, Sint and Srel); the English tags S (simple declarative clause), SBAR (clause introduced by a subordinating conjunction), SBARQ (direct question introduced by a wh-word or a wh-phrase), SINV (declarative sentence with subject-aux inversion), SQ (subconstituent of SBARQ excluding wh-phrase), PRN (parenthetical), FRAG (fragment) and RRC (reduced relative clause). The English tag FRAG marks those portions of text that appear to be clauses but lack too many essential elements for the exact structure to be easily determined; the RRC label is used only in cases where "there is no VP and an extra level is needed for proper attachment of sentential modifiers" [7]. FRAG and RRC tags are at the same level as SBAR, SINV, and SQ, so they are classified uniformly into S category in the mapping, for instance:

(FRAG (, ,)
    (NP (JJ for) (NN example)))
(RRC

```
(ADVP (RB now)
 (ADJP
  (NP
   (QP (JJR more) (IN than) (CD 100))
   (NNS years))
  (JJ old))))
```

The English tag PRN (parenthetical) is used to tag the parenthetical contents which are usually like sub-sentence and there is no such tag in the French phrase tagset, so PRN is also considered as under the category of sub-sentence in the universal tags, for instance:

```
(PRN (, ,)
 (S
  (NP (PRP it))
  (VP (VBZ seems)))
 (, ,))
(PRN (-LRB- -LRB-)
 (NP
  (NP (CD 204) (NNS boys))
  (CC and)
  (NP (CD 172) (NNS girls)))
 (PP (IN from)
  (NP (CD 13) (NNS schools)))
 (-RRB- -RRB-))
```

The universal phrase tag COP (coordinated phrase) covers: the French tag COORD (coordinated phrases); the English tag UCP (unlike coordinated phrase, labeling the coordinated phrases that belong to different categories).

The universal phrase tag CONJP (conjunction phrase) covers: the English tag CONJP (conjunction phrase). The French Treebank has no homologous tags that have the similar function to the Penn Treebank phrase tag CONJP (conjunction phrase), for instance:

```
(CONJP (RB as) (RB well) (IN as))
```

The universal phrase tag X (other phrases or unknown) covers: the English tags X (unknown or uncertain), INTJ (interjection), LST (list marker).

3   Application in MT Evaluation

The conventional MT evaluation methods include the measuring of the closeness of word sequences between the translated documents (hypothesis translation) and the reference translations (usually offered by professional translators), such as the BLEU [8], NIST [9] and METEOR [10], and the measuring of the editing distance from the hypothesis translation to the reference translations, such as the TER [11]. However, the reference translations are usually expensive. Furthermore, the reference translations are in fact not available when the translation task contains a large amount of data. The phrase tags mapping designed for French and English treebanks is employed to develop a novel evaluation approach for French-to-English MT translation without using the money-and-time-consuming reference translations. The evaluation is performed on the French (source) and English (target) documents.

3.1   Designed Methods

We assume that the translated (targeted) sentence should have a similar set of phrase categories with the source sentence. For instance, if the source French sentence has an adjective phrase and a prepositional phrase then the targeted English sentence is most likely to have an adjective phrase and a prepositional phrase, too. This design is inspired by the synonymous relation between the source sentence and the target sentence. Admittedly, it is possible that we get two sentences that have the similar set of phrase categories but talk about different things. However, this evaluation approach is not designed for the general circumstance. This evaluation approach is designed and applied under the assumption that the targeted sentences are indeed the translated sentences (usually by some automatic MT systems) from the source document. Under this assumption, if the translated sentence retains the complete meaning of the source sentence, then it usually contains most (if not all) of the phrase categories in the source sentence. Furthermore, the Spearman correlation coefficient will be employed in the experiments to test this evaluation approach (to measure the correlation scores of this method derived by comparing with the human judgments). Following the common practice, human judgments are assumed as the Gold Standard [12][13].

Assume we have the following two sentences: sentence X is the source French and sentence Y is the targeted English (translated by a MT system). Firstly, the two sentences are parsed by the corresponding language parsers respectively (this paper uses the Berkeley parsers [14]). Then the phrase tags of each sentence are extracted from the parsing results. Finally, the quality estimation of the translated sentence Y is conducted based on these two sequences of extracted phrase tags. See Figure 1 and Figure 2 for explanation (PhS and UniPhs represent phrase sequence and the universal type of phrase sequence respectively). To make it both concise and content-rich, the phrase tag at the second level of each word and punctuation (counted from bottom up, namely the tokens in the parsing tree, i.e. just the immediate level above the POS tag level) is extracted. Thus, the extracted phrase sequence (PhS) has the same length as that of the parsed sentence.

3.2   Designed Formula

The precision and recall values are conventionally regarded as the important factors to reflect the quality of system performances in NLP literature. For instance, the BLEU metric employs uniformly weighted 4-gram precision scores and METEOR employs both unigram precision and unigram recall combined with synonym and stem dictionaries. This paper designs a formula that contains

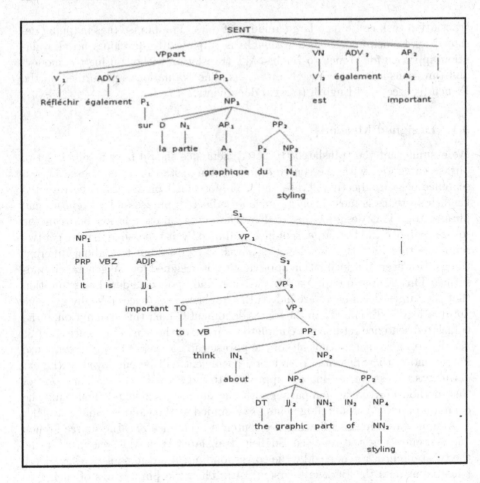

Fig. 1. The parsing results of French and English sentences

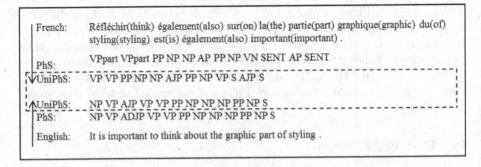

French:	Réfléchir(think) également(also) sur(on) la(the) partie(part) graphique(graphic) du(of) styling(styling) est(is) également(also) important(important) .
PhS:	VPpart VPpart PP NP NP AP PP NP VN SENT AP SENT
∇UniPhS:	VP VP PP NP NP AJP PP NP VP S AJP S
△UniPhS:	NP VP AJP VP VP PP NP NP NP PP NP S
PhS:	NP VP ADJP VP VP PP NP NP NP PP NP S
English:	It is important to think about the graphic part of styling .

Fig. 2. The converting of extracted phrase tags into universal categories

weighted N_2 gram precision, weighted N_3 gram recall and N_1 gram position order factors. Different n-gram precisions and recalls are combined respectively with weighted geometric means (using different weight parameters, different with BLEU). The whole formula is grouped using the weighted harmonic mean. We design a factor, the position difference, due to the fact that there is indeed word order information within the phrase. However, the request on the sequence order becomes weak at phrase levels (e.g. some languages have free phrase order). According to the analysis above, the weight of position factor will be adjusted to be lower than the weights of precision and recall.

$$HPPR = Har(w_{Ps}\overline{N_1PsDif}, w_{Pr}N_2Pre, w_{Rc}N_3Rec)$$

$$= \frac{w_{Ps} + w_{Pr} + w_{Rc}}{\frac{w_{Ps}}{\overline{N_1PsDif}} + \frac{w_{Pr}}{N_2Pre} + \frac{w_{Rc}}{N_3Rec}} \tag{1}$$

In the formula, Har means harmonic mean; the variables, $\overline{N_1PsDif}$, N_2Pre and N_3Rec, represent N_1 gram position difference factor, weighted N_2 gram precision and weighted N_3 gram recall (all at corpus level) respectively. The parameters w_{Ps}, w_{Pr} and w_{Rc} are the weights of the corresponding three variables $\overline{N_1PsDif}$, N_2Pre and N_3Rec. Let's see the introduction of the sentence-level N_1PsDif firstly.

$$N_1PsDif = e^{-N_1PD} \tag{2}$$

$$N_1PD = \frac{1}{Length_{hyp}} \sum_{i=1}^{Length_{hyp}} |PD_i| \tag{3}$$

$$|PD_i| = |PsN_{hyp} - MatchPsN_{src}| \tag{4}$$

The variables, N_1PD, $Length_{hyp}$ and PD_i, mean N_1 gram position difference value of a sentence, hypothesis sentence length and position difference value of the ith tag in the hypothesis sentence respectively. The variables PsN_{hyp} and $MatchPsN_{src}$ mean the position number of a tag in the hypothesis sentence and the corresponding position number of the matching tag in the source sentence. If there is no successful match for the hypothesis tag, then:

$$|PD_i| = |PsN_{hyp} - 0| \tag{5}$$

To calculate the value of N_1PsDif, we should first finish the N_1 gram UniPhS alignment from the hypothesis to the source sentence (direction fixed). In this case, the n-gram alignment means that the candidate matches that also have neighbor matches (neighbors are limited to the previous and following N_1 tags) will be assigned higher priority. If there is no surrounding match for all potential tags then the nearest match operation will be selected as a backup choice. On the other hand, if there are more than two candidates in the source sentence that both have the neighbor matching with the hypothesis tag, then the candidate

```
for x = 1 ⋯ l_h
    for y = 1 ⋯ l_s
        if unmatch(tag_x, tag_y^s)
            Align(tag_x, ∅);
        elseif match(tag_x, tag_y^s)
            Align(tag_x, tag_y^s);
        else          // more than one candidates matching tag_x (i ≥ 2)
            LookforFinalAlign(tag_x, < tag_{y_1}^s, …, tag_{y_i}^s >)
    endfor
endfor

LookforFinalAlign(tag_x, < tag_{y_1}^s, …, tag_{y_i}^s >)
for p = -N_1, …, -1,1, ⋯, N_1
    for q = -N_1, …, -1,1, ⋯, N_1
        if unmatch(tag_{x+p}, < tag_{y_1+q}^s, …, tag_{y_i+q}^s >)  //no candidate tag has neighbor match
            Align(tag_x, tag_{y_j}^s) when y_j (1 < j < i) has the shortest distance with x;
        elseif match(tag_{x+p}, tag_{y_j+q}^s)   //only tag_{y_j}^s has neighbor match with tag_x
            Align(tag_x, tag_{y_j}^s);
        else          // more than two candidates have neighbor match with tag_x
            Align(tag_x, tag_{y_j}^s) when y_j (1 < j < i) has the shortest distance with x;
    endfor
endfor
```

Fig. 3. The N_1 gram alignment algorithm

that has the nearest match with the hypothesis tag will also be selected. The final result is restricted to one-to-one alignment. The alignment algorithm is shown in Figure 3. The variable l_h and l_s represent the length of hypothesis sentence and source sentence respectively. The tag_x and tag_y^s mean the phrase tag in hypothesis and source sentence respectively. In this paper, the N_1gram is performed using bigram (N_1=2). After the alignment, an example for the N_1PD calculation of a sentence is shown in Figure 4. Then the value of the N_1 gram Position difference factor $\overline{N_1PsDif}$ at the corpus level is the arithmetical mean of each sentence level score N_1PsDif_i.

$$\overline{N_1PsDif} = \frac{1}{n}\sum_{i=1}^{n} N_1PsDif_i \qquad (6)$$

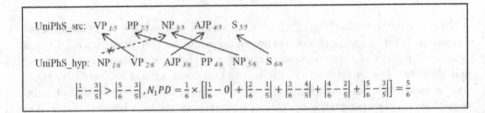

Fig. 4. The N_1PD calculation example

The values of the corpus-level weighted N_2 gram precision N_2Pre and weighted N_3 gram recall N_3Rec are calculated as follows:

$$N_2Pre = exp(\sum_{n=1}^{N_2} w_n log P_n) \tag{7}$$

$$N_3Rec = exp(\sum_{n=1}^{N_3} w_n log R_n) \tag{8}$$

The variables P_n and R_n mean the n-gram precision and n-gram recall that are calculated as follows:

$$P_n = \frac{\#matched\ ngram\ chunks}{\#ngram\ chunks\ of\ hypothesis\ corpus} \tag{9}$$

$$R_n = \frac{\#matched\ ngram\ chunks}{\#ngram\ chunks\ of\ source\ corpus} \tag{10}$$

The variable $\#matched\ ngram\ chunks$ represents the number of matched n-gram chunks between the hypothesis and source corpus (sum of sentence-level matches). Each n-gram chunk could be used only once, and an example is the bigram chunks matching in Figure 5.

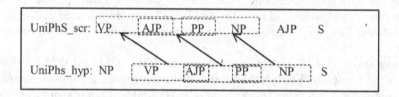

Fig. 5. Bigram chunks matching example

To calculate n-gram precision and recall values, we should know that if they are designed at the sentence level then some of the sentence-level bigram or trigram precision and recall values will be 0 due to the fact that some hypothesis sentences result in no bigram or trigram chunk matching with the source sentences (it is meaningless when the variable of logarithmic function is less than or equal to 0). According to the analysis above, after the sentence-level n-gram chunks matching, we calculate the n-gram precision P_n and n-gram recall R_n values at the corpus level as shown in the formula (this design is similar to BLEU; BLEU calculates the corpus level n-gram precisions). Thus the geometric mean of weighted N_2 gram precision N_2Pre and weighted N_3 gram recall N_3Rec will also be at the corpus level.

4   Experiments

4.1   Evaluating Method for Evaluation

Spearman rank correlation coefficient (rs) is conventionally used as the evaluating method for the MT evaluation metrics [12][13]. The evaluation result that has a higher correlation score, namely higher correlation with human judgments, will be regarded as having a better quality.

$$rs_{XY} = 1 - \frac{6 \sum_{i=1}^{n} d_i^2}{n(n^2 - 1)} \tag{11}$$

We assume two sequence $\vec{X} = \{x_1, x_2, ..., x_n\}$ and $\vec{Y} = \{y_1, y_2, ..., y_n\}$. Firstly we get the corresponding rank sequences $\hat{\vec{X}} = \{\hat{x_1}, \hat{x_2}, ..., \hat{x_n}\}$ and $\hat{\vec{Y}} = \{\hat{y_1}, \hat{y_2}, ..., \hat{y_n}\}$; then d_i is the difference-value between the coordinate variables, i.e. $\hat{x_i} - \hat{y_i}$.

4.2   Experiments and Results

The corpora used in the experiments are from the international workshop of statistical machine translation (WMT) which has been held annually by the ACL's special interest group of machine translation (SIGMT[1]). To avoid the overfitting problem, the WMT 2011 corpora are used as the development set (for tuning the weights of factors in the formula to make the evaluation results close to the human judgments). Then the WMT 2012 corpora are used as the testing set with the formula that has the same parameters tuned in the development stage.

There are 18 and 15 systems respectively in WMT 2011 and WMT 2012 producing the French-to-English translation documents, each document containing 3003 sentences. Each year, there are hundreds of human annotators to evaluate the MT system outputs, and the human judgments task usually costs hundreds of hours of labor. The human judgments are used to validate the automatic metrics. The system-level Spearman correlation coefficient of the different evaluation results will be calculated as compared to the human judgments. The state-of-the-art evaluation metrics BLEU (measuring the closeness between the hypothesis and reference translations) and TER (measuring the editing distance) are selected for the comparison with the designed model $HPPR$.

The values of N_2 and N_3 are both selected as 3 due to the fact that the 4-gram chunk match usually results in 0 score. The tuned factor weights in the formula are shown in Table 2. The experiment results on the development and testing corpora are shown in Table 3 and Table 4 respectively ("Use Reference?" means whether this metric uses reference translations). The experiment results on the development and testing corpora show that $HPPR$ without using reference translations has yielded promising correlation scores (0.63 and 0.59 respectively) with human judgments even though lower than the reference-aware

---

[1] http://www.sigmt.org/

Table 2. Tuned factor weights

Items	Weight parameters	Ratio
N_2Pre	$w_1 : w_2 : w_3$	8:1:1
N_3Rec	$w_1 : w_2 : w_3$	1:1:8
$HPPR$	$w_{Ps} : w_{Pr} : w_{Rc}$	1:8:1

Table 3. Development results on WMT 2011 corpus

Metrics	Use Reference?	Spearman rs
BLEU	Yes	0.85
TER	Yes	0.77
$HPPR$	No	0.63

metrics BLEU and TER. Results also specify that there is still potential to improve the performances of all the three metrics, even though that the correlation scores which are higher than 0.5 are already considered as strong correlation [15] as shown in Table 5.

5 Discussion

To facilitate future research in multilingual or cross-lingual literature, this paper designs a phrase tags mapping between the French Treebank and the English Penn Treebank using 9 commonly used phrase categories. One of the potential applications in the MT evaluation task is explored in the experiment section. This novel evaluation model could spare the reference translation corpora that are conventionally used in the MT evaluation tasks. This evaluation approach is favorable especially for the circumstance where there are many translation systems outputting a lot of translated documents and the reference translations are very expensive or not easily available in time.

There are still some limitations in this work to be addressed in the future. First, the designed universal phrase categories may not be able to cover all the phrase tags of other language treebanks, so this tagset could be expanded when necessary. Second, in the application of MT evaluation of the mapping work, the designed formula contains the n-gram factors of position difference, precision and recall, which may not be sufficient or suitable for some of the other

Table 4. Testing results on WMT 2012 corpus

Metrics	Use Reference?	Spearman rs
BLEU	Yes	0.81
TER	Yes	0.82
$HPPR$	No	0.59

Table 5. Correlation scores analysis

Correlation	Negative	Positive
None	-0.09 to 0	0 to 0.09
Small	-0.3 to -0.1	0.1 to 0.3
Medium	-0.5 to -0.3	0.3 to 0.5
Strong	-1.0 to -0.5	**0.5 to 1.0**

language pairs, so different measuring factors should be added or switched when facing new challenges.

Acknowledgments. The authors wish to thank the anonymous reviewers for many helpful comments and Mr. Yunsheng Xie who is a Master Candidate of Arts in English Studies from the Department of English in the Faculty of Social Sciences and Humanities at the University of Macau for his kind help in improving the language quality.

References

1. Marcus, M., Santorini, B., Marcinkiewicz, M.A.: Building a large annotated corpus of English: The Penn Treebank. Computational Linguistics 19(2), 313–330 (1993)
2. Mitchell, M., Kim, G., Marcinkiewicz, M.A., MacIntyre, R., Bies, A., Ferguson, M., Katz, K., Schasberger, B.: The Penn Treebank: Annotating Predicate Argument Structure. In: Human Language Technology: Proceedings of Workshop, Plainsboro, New Jersey, March 8-11, pp. 114–119. H94-1020 (1994)
3. Skut, W., Krenn, B., Brants, T., Uszkoreit, H.: An annotation scheme for free word order languages. In: Proc. of ANLP, pp. 88–95 (1997)
4. Abeillé, A., Clément, L., Toussenel, F.: Building a Treebank for French. Building and Using Parsed Corpora. In: Abeillé (Abeillé, 2003), ch. 10. ANNE Abeillé, Treebanks. Kluwer Academic Publishers (2003)
5. Chen, K., Luo, C., Chang, M., Chen, F., Chen, C., Huang, C., Gao, Z.: Sinica tree-bank: Design criteria, representational issues and implementation. In: Abeillé, ch. 13, pp. 231–248 (2003)
6. Slav, P., Das, D., McDonald, R.: A Universal Part-of-Speech Tagset. In: Proceedings of the Eight International Conference on Language Resources and Evaluation (LREC 2012), Istanbul, Turkey (2012)
7. Bies, A., Ferguson, M., Katz, K., MacIntyre, R.: Bracketing Guidelines for Treebank II style Penn Treebank Project. Linguistic Data Consortium (1995)
8. Kishore, P., Roukos, S., Ward, T., Zhu, W.-J.: BLEU: a method for automatic evaluation of machine translation. In: Proceedings of the 40th Annual Meeting on Association for Computational Linguistics (ACL 2002), pp. 311–318. Association for Computational Linguistics, Stroudsburg (2002)
9. George, D.: Automatic evaluation of machine translation quality using n-gram co-occurrence statistics. In: Proceedings of the Second International Conference on Human Language Technology Research (HLT 2002), pp. 138–145. Morgan Kaufmann Publishers Inc., San Francisco (2002)

10. Satanjeev, B., Lavie, A.: METEOR: An Automatic Metric for MT Eval-uation with Improved Correlation with Human Judgments. In: Proceedings of the 43th An-nual Meeting of the Association of Computational Linguistics (ACL 2005), pp. 65–72. Association of Computational Linguistics, Ann Arbor (June 2005)
11. Matthew, S., Dorr, B.J., Schwartz, R., Micciulla, L., Makhoul, J.: A Study of Trans-lation Edit Rate with Targeted Human Annotation. In: Proceedings of the 7th Con-ference of the Association for Machine Translation in the Americas (AMTA 2006), USA, pp. 223–231 (2006)
12. Callison-Burch, C., Koehn, P., Monz, C., Zaidan, O.F.: Findings of the 2011 Workshop on Statistical Machine Translation. In: Proceedings of the Sixth Work-shop on Statistical Machine Translation of the Association for Computational Linguistics(ACL-WMT), pp. 22–64. Association for Computational Linguistics, Edinburgh (2011)
13. Chris, C.-B., Koehn, P., Monz, C., Post, M., Soricut, R., Specia, L.: Findings of the 2012Workshop on Statistical Machine Translation. In: Pro-ceedings of the Seventh Workshop on Statistical Machine Translation, pp. 10–51. Association for Computational Linguistics, Mon-treal (2012)
14. Petrov, S., Barrett, L., Thibaux, R., Klein, D.: Learning accurate, compact, and interpretable tree annotation. In: Proceedings of the 21st International Con-ference on Computational Linguistics and the 44th Annual Meeting of the Association for Computational Linguistics (ACL-44), pp. 433–440. Association for Computational Linguistics, Strouds-burg (2006)
15. Cohen, J.: Statistical power analysis for the behavioral sciences, 2nd edn. Psychol-ogy Press (1988)

Statistical Machine Translation of Subtitles:
From OpenSubtitles to TED

Mathias Müller and Martin Volk

Institute of Computational Linguistics, Zurich, Switzerland

Abstract. In this paper, we describe how the differences between subtitle corpora, OpenSubtitles and TED, influence machine translation quality. In particular, we investigate whether statistical machine translation systems built on their basis can be used interchangeably. Our results show that OpenSubtiles and TED contain very different kinds of subtitles that warrant a subclassification of the genre. In addition, we have taken a closer look at the translation of questions as a sentence type with special word order. Interestingly, we found the BLEU scores for questions to be higher than for random sentences.

1 Introduction

The key ingredient for building successful Statistical Machine Translation (SMT) systems is a suitable and sufficiently large parallel corpus. For a number of language pairs, large subtitle corpora are available. The OPUS OpenSubtitles corpus (Tiedemann, 2009) contains fansubs for 54 languages. This collection amounts to several million parallel sentences for the most popular language pairs. On the other hand, there is the collection of subtitles from the TED talks (Cettolo et al., 2012), a series of high quality talks on "Technology, Entertainment, and Design". Although the TED collection is much smaller, it is interesting because of its wide coverage and complexity of topics.

Therefore, we set out to investigate the degree of similarity between the subtitles in the two corpora and to what extent this influences the quality of SMT systems trained on them. As a side issue we wanted to check the general usefulness of the OpenSubtitles corpus for training SMT systems which has been questioned repeatedly in the literature (see e.g. (Petukhova et al., 2012)). In order to gain deeper insight into the impact of the two corpora on specific linguistic phenomena, we evaluated their translation quality on questions.

In this paper we describe the differences between the two subtitle corpora and their impact on translation quality. We trained a number of Moses systems for that purpose, using parallel subtitles in English, French and German. In all, we trained twelve systems under the exact same conditions (preprocessing, Moses command line options). Each system was then tested on three test sets including a set containing only questions.

I. Gurevych, C. Biemann, and T. Zesch (Eds.): GSCL 2013, LNAI 8105, pp. 132–138, 2013.
© Springer-Verlag Berlin Heidelberg 2013

2   OpenSubtitles vs. TED

Both OpenSubtitles and TED are collections of parallel sentences derived from subtitles. So, we are dealing with sentence-aligned corpora and not subtitle-aligned corpora as, for instance, Volk et al. (2010). Contrary to our intuition, Volk et al. (2010) found that subtitle-aligned corpora are as good for building SMT systems for subtitles as sentence-aligned corpora. The average sentence length is around 6 words per sentence for OpenSubtitles and around 17 for TED, irrespective of the language. It is surprising that the figures across languages are so similar. We believe that this is an artifact of the automatic alignment. Only those sentences that are of similar length were aligned. However, these numbers clearly indicate that OpenSubtitles and TED subtitles are different from one another. Some randomly chosen lines may illustrate this point.

Example 1. **Subtitle examples from OpenSubtitles and TED**[1]
OpenSubs: You miss me, today?
OpenSubs: Faut y penser avant.

TED: The first law is two-colorability: You can color any crease pattern with just two colors without ever having the same color meeting.
TED: Et avec ce qu'il trouve sur place, il entre et fait son petit studio qui sert de base de travail.

The differences in length can be explained if we consider the circumstances: OpenSubtitles are taken from regular movies and series, where sentences tend to be short and where subtitles are shortened to fit on the screen in the available time. TED talks on the other hand treat rather complex subjects that in turn demand more complex sentences. Any differences may have an influence on MT performance when crosstesting between corpora, more precisely on BLEU scores in our case. For our experiments, we used OpenSubtitles, as prepared in the OPUS project, for language pairs between EN, FR and DE in all possible combinations. The OpenSubtitles collections for these languages are large, ranging from 2.8 million sentence pairs for DE-FR up to 20 million sentence pairs for EN-FR. These corpora should be sufficient, given that Hardmeier and Volk (2009) argue that 1 million sentence pairs is suitable for subtitle SMT. Still we need to remember that subtitles in OpenSubtitles are of an unknown quality. Some are controlled and consistent, but others contain spelling errors or strange wordings. All texts are already sentence-aligned and formatted in a Moses-friendly way. As for the limitations of sentence-alignment techniques, see Tiedemann (2009).

The TED collection (Cettolo et al., 2012) is smaller. These subtitles are crawled from www.ted.com, a platform offering talks that were recorded at TED conferences or similar events. These videos sometimes come with subtitles translated into 30 or more languages by volunteers from within the TED community. The translations are generally high-quality because TED requires translators

---

[1] Obviously unrelated sentences, not translations that correspond.

to peer-review their work and prove their proficiency. These corpora, too, have been preprocessed and sentence-aligned by (Cettolo et al., 2012) as to allow using them in Moses with ease. Again, we only used the parts in EN, FR and DE.

In total, we built 12 SMT systems between EN, FR and DE on the basis of 6 different corpora from TED and OpenSubtitles. We used the same material for both directions, for example the same corpus to translate EN-FR and FR-EN. Table 1 shows the corpora's sizes as the number of lines (roughly equal to the number of sentences) and the number of words using a naive tokenization.

Table 1. Corpus sizes

Corpus	Language	Number of lines	Number of words
OpenSubtitles EN-DE (DE-EN)	EN	4,654,635	26,266,191
	DE	"	27,189,072
OpenSubtitles EN-FR (FR-EN)	EN	19,858,798	119,682,551
	FR	"	115,456,439
OpenSubtitles DE-FR (FR-DE)	DE	2,862,370	16,946,049
	FR	"	16,818,332
TED EN-DE (DE-EN)	EN	63,865	1,029,090
	DE	"	1,034,657
TED EN-FR (FR-EN)	EN	114,582	1,916,788
	FR	"	2,000,958
TED DE-FR (FR-DE)	DE	62,148	967,935
	FR	"	1,056,758

3   Building SMT Systems

Starting out with the corpora described in the section above, we built phrase-based Moses systems (Koehn et al., 2007) and tested their performance. Moses is used widely and is the state-of-the-art tool for statistical machine translation. We divided each corpus into training set (97 percent), tuning set (1 percent) and test set (2 percent) and assigned parallel lines randomly to the sets[2]. All of the data passed through the usual stages of preprocessing as we cleaned, lowercased and tokenized it using scripts offered by the Moses toolkit. For word-alignment we used GIZA++ (Och and Ney, 2003) which is implemented as part of Moses.

In order to build the language model we employed SRILM (Stolcke et al., 2011) together with Kneser-Ney discounting for smoothing, and interpolation as a back-off model for probabilities. These two options are an official recommendation by the Moses developers[3]. In general, we used the standard methods and options where possible and consistently applied the same rules to all systems.

---

[2] While creating the test set via randomly allocating the lines is statistically sound, it might be more natural to test the subtitles of whole movies or series.

[3] See http://www.statmt.org/moses/?n=FactoredTraining/BuildingLanguageModel for further information.

The baseline systems were then tuned with MERT (also part of Moses), optimizing them with respect to the tuning set. After assembling complete Moses systems, we tested each on three different test sets to obtain BLEU scores.

4 First Results

We conducted several experiments on the TED and OpenSubtitles collections. In all cases, the performance of the MT systems was measured with *multibleu*, a script distributed with Moses. Table 2 reports the BLEU scores of our systems. The most straightforward test set for each system is the 2 percent of the original corpus set aside at the beginning, the "native test set". In contrast, the "foreign test set" is the native set's equivalent from the other collection. In other words, TED systems are subjected to foreign test data taken from OpenSubtitles and vice versa.

Table 2. Performance results in BLEU scores

Language pair	System	Test set	
		native	foreign
DE-EN	OpenSubtitles	27.92	20.56
	TED	25.06	14.29
DE-FR	OpenSubtitles	17.18	14.65
	TED	17.64	9.69
EN-DE	OpenSubtitles	19.55	16.93
	TED	24.38	12.59
EN-FR	OpenSubtitles	22.86	23.56
	TED	31.87	14.89
FR-DE	OpenSubtitles	13.42	10.72
	TED	13.12	8.34
FR-EN	OpenSubtitles	23.52	**24.92**
	TED	**33.37**	16.87

First, let us consider the difference in performance between OpenSubtitles and TED systems when confronted with their native test sets. Out of six OpenSubtitles systems, the highest scores are achieved with DE-EN, scoring almost 28. Only TED systems surpass 30, and only when translating between EN and FR. Thus, good performance results wherever EN is involved, irrespective of its being the source or target language. On the other hand, combinations of DE and FR lead to the lowest scores. To understand this, we have to bear in mind that most movies and all TED talks are in English in the first place. Often, the English transcription is done first and translators base their work on English subtitles. Therefore, combinations with English (EN-FR, FR-EN, DE-EN, EN-DE) can be expected to be translations of one another, whereas combinations between DE and FR (DE-FR, FR-DE) are not directly related.

With regard to the foreign sets, the performance ranking changes somewhat. OpenSubtitles translating from FR to EN now takes the lead, resulting in a BLEU score of approximately 25. In general, the TED systems are affected more severely when confronted with foreign test lines, their scores plummeting to 50 percent of the former value in some cases.This indicates that the OpenSubtitles systems are more apt at translating TED than the other way round. Also, it implies that TED systems are more overfitted and OpenSubtitles systems more universal if our goal is to translate subtitles in general.

5   Investigating MT Quality for Questions

As a case study we have investigated the MT quality of questions. Questions are special because they have word order that is different from assertive clauses, and they use question words and special auxiliary verbs. For the sake of simplicity, a question is a line ending with a question mark. Here are some typical questions.

Example 2. **Question examples from OpenSubtitles and TED**
OpenSubs: Is the needle in his femoral artery, Mr. Palmer?
OpenSubs: Für dich sind wir nur Leichen, oder?

TED: And we asked ourselves, why couldn't it be exhibitionistic, like the Met, or like some of the other buildings at Lincoln Center?
TED: Was ist die Botschaft, was ist das Vokabular und die Grammatik, die von diesem Gebäude ausgesandt wird, und was sagt es uns über uns selbst?

A line that ends in a question mark in some cases might not be a question. For example, the English TED sentence in the example above is an assertion or an indirect question. But such cases are rare and are ignored here. Some of the lines ending with a question mark in one language do not have an equivalent counterpart, i.e. in the translation there is no question mark at the end. We disregarded them for our tests.

Our questions test set contains only questions from the native test set. We used the "question set" both for a quantitative (performance measures in BLEU scores) and qualitative analysis (manual error categorisation). With respect to questions, OpenSubtitles systems performed slightly better compared to the native test set, their scores climbing one or two BLEU points (see table 3). The scores of TED systems adapted slightly. Given the differences in performance between the native and question set, questions surprisingly score higher than the average subtitle of any type. We speculate that this might be due to the fact that questions are shorter than the average subtitle. The latter is 33.7 characters long – calculated over all the lines we took from OpenSubtitles. The average question line taken from OpenSubtitles counts no more than 27.2 characters. We get similar values for the TED corpora.

In order to evaluate the performance of translating questions qualitatively, we have looked at up to 100 translated questions from EN-DE and FR-DE, both

Table 3. BLEU scores for native and questions test sets

System	Test set	
	native	questions
OpenSubtitles DE-EN	27.92	31.11
TED DE-EN	25.06	27.09
OpenSubtitles DE-FR	17.18	19.61
TED DE-FR	17.64	17.14
OpenSubtitles EN-DE	19.55	21.73
TED EN-DE	24.38	27.99
OpenSubtitles EN-FR	22.86	23.28
TED EN-FR	31.87	29.7
OpenSubtitles FR-DE	13.42	15.37
TED FR-DE	13.12	15.37
OpenSubtitles FR-EN	23.52	23.55
TED FR-EN	33.37	30.61

OpenSubs and TED. In particular, we have paid attention to the types of errors that occur. Our categories are fragmentation (translation unit span too narrow), omission, lack of agreement, difficulty with ambiguous terms, reordering, ulexis issues and addition (of a phrase). The following errors were repeatedly made. Systems translating from FR to DE frequently omitted an infinitive, whereas this never happened when translating from EN to DE. Also, only EN-DE systems treated many auxiliary verbs as full verbs. Out-of-vocabulary problems are a more serious issue with FR-DE, presumably because verb-pronoun compounds like "atterrissez-vous" or "a-t-il" are common in French. The majority of lexis errors is concerned with a hyphenated French word like those. Deliberately tokenizing these forms as part of the preprocessing would alleviate this effect.

6   Conclusion

Parallel corpora of subtitles are a valuable source for machine translation and are frequently used. We compared corpora from the TED and OpenSubtitles collections, and we suggest that "subtitles" is in fact too broad a category. Four rows of test sets revealed that the systems can hardly be used interchangeably, since sentence length, broad applicability and subtitle quality mark stark differences between subtitles from OpenSubtitles and from TED. They may be so different that they might best be treated as different genres indeed. We isolated questions and found slightly better BLEU scores for them as compared to randomly selected sentences.

In future studies, it might prove fruitful to incorporate data from both collections into one system and assign weights to each in order to counteract the different sizes of the training corpora. One way to achieve this is to combine the phrase tables resulting from building translation models (see Sennrich (2012)).

References

Cettolo, M., Girardi, C., Federico, M.: Wit[3]: Web inventory of transcribed and translated talks. In: Proceedings of the 16th Conference of the European Association for Machine Translation (EAMT), Trento, Italy, pp. 261–268 (2012)

Hardmeier, C., Volk, M.: Using linguistic annotations in statistical machine translation of film subtitles. In: Nodalida (2009)

Koehn, P., Hoang, H., Birch, A., Callison-Burch, C., Federico, M., Bertoldi, N., Cowan, B., Shen, W., Moran, C., Zens, R., Dyer, C., Bojar, O., Constantin, A., Herbst, E.: Moses: Open source toolkit for statistical machine translation. In: ACL (2007)

Och, F.J., Ney, H.: A systematic comparison of various statistical alignment models. Computational Linguistics 29(1), 19–51 (2003)

Petukhova, V., Agerri, R., Fishel, M., Penkale, S., del Pozo, A., Maucec, M.S., Way, A., Georgakopoulou, P., Volk, M.: SUMAT: Data collection and parallel corpus compilation for machine translation of subtitles. In: LREC, pp. 21–28 (2012)

Sennrich, R.: Perplexity minimization for translation model domain adaptation in statistical machine translation. In: Proceedings of the 13th Conference of the European Chapter of the Association for Computational Linguistics (EACL), Avignon, France (2012)

Stolcke, A., Zheng, J., Wang, W., Abrash, V.: SRILM at sixteen: Update and outlook. In: Proceedings IEEE Automatic Speech Recognition and Understanding Workshop (2011)

Tiedemann, J.: News from opus - a collection of multilingual parallel corpora with tools and interfaces. In: Nicolov, N., Bontcheva, K., Angelova, G., Mitkov, R. (eds.) Recent Advances in Natural Language Processing, vol. V, pp. 237–248. John Benjamins, Amsterdam (2009)

Volk, M., Sennrich, R., Hardmeier, C., Tidström, F.: Machine translation of TV subtitles for large scale production. In: Proceedings of the Second Joint EM+/CNGL Workshop on Bringing MT to the User: Research on Integrating MT in the Translation Industry, Denver, pp. 53–62 (2010)

Part-Of-Speech Tagging for Social Media Texts

Melanie Neunerdt[1], Bianka Trevisan[2], Michael Reyer[1], and Rudolf Mathar[1]

[1] Institute for Theoretical Information Technology,
[2] Textlinguistics/TechnicalCommunications,
RWTH Aachen University, Germany

Abstract. Work on Part-of-Speech (POS) tagging has mainly concentrated on standardized texts for many years. However, the interest in automatic evaluation of social media texts is growing considerably. As the nature of social media texts is clearly different from standardized texts, Natural Language Processing methods need to be adapted for reliable processing. The basis for such an adaption is a reliably tagged social media text training corpus. In this paper, we introduce a new social media text corpus and evaluate different state-of-the-art POS taggers that are retrained on that corpus. In particular, the applicability of a tagger trained on a specific social media text type to other types, such as chat messages or blog comments, is studied. We show that retraining the taggers on in-domain training data increases the tagging accuracies by more than five percentage points.

Keywords: POS tagging, statistical NLP, social media texts.

1 Introduction

Many Natural Language Processing (NLP) methods, e.g., syntactical parsing or sentiment analysis, require accurate Part-of-Speech (POS) tag information for a given word sequence. This information is provided by automatic POS tagging which is a well researched field. For German, which has a strong morphological character, state-of-the art POS taggers yield per-word accuracies of 97% to 98% for standardized texts.

However, the interest in using NLP methods for non-standardized texts, such as social media texts, is growing. The automatic evaluation of social media texts is particularly essential for the task of sentiment analysis. Social media texts comprise user generated content such as blog comments or chat messages. Indeed, differ from standardized texts in the word usage but also in their grammatical structure. This holds for adapted NLP methods in general and in particular for the adaption of POS tagging methods to such text types. Most state-of-the-art taggers are developed for standardized texts. Hence they are trained on large newspaper corpora. However, previous studies, e.g., [6], have shown that applying such taggers to non-standardized texts results in a significant performance loss. The lack of social media text reference corpus, which is sufficiently large to train a tagger, might be the reason that automatic POS tagging for social media texts

I. Gurevych, C. Biemann, and T. Zesch (Eds.): GSCL 2013, LNAI 8105, pp. 139–150, 2013.
© Springer-Verlag Berlin Heidelberg 2013

has rarely been studied so far. Furthermore it has not been addressed for German language, yet. Hence, tagging methods need to be adapted and annotation rules for social media text characteristics are required. A first step is to provide in-domain training data to yield higher tagging accuracies when existing taggers are applied to social media texts.

In this paper, we first introduce a new social media text corpus containing Web comments composed of 36,000 annotated tokens. Considering the German standard *Stuttgart/Tübinger TagSet* (STTS) [11], we distinguish 54 tag classes. Until now, no standardized STTS extension has been published for the annotation of social media texts in German. Thus, we define annotation rules for specific social media text characteristics.

We use the introduced corpus to retrain existing taggers. Combining the social media text corpus with standardized texts (joint-domain), four state-of-the-art taggers, TreeTagger [13], TnT [3], Stanford [14], and SVMTool [7], are trained and evaluated by cross validation. We show that tagging accuracies increase by more than five percentage points for social media texts. The *TIGER* corpus [2] serves as standardized training data in the experiments. Mean tagging accuracies with standard deviations are calculated and compared for all taggers. A more detailed look into the results is provided for the TreeTagger, which has the highest tagging accuracies on German social media texts.

Finally, we study the applicability of the retrained taggers to four different social media text types, i.e., blog comments, chat messages, YouTube comments and news site comments. For the evaluation, 5,000 tokens that comprise the four types are additionally manually annotated with POS tags by manual processing. In particular, tagging errors are classified into four categories, with respect to different social media text characteristics. This particularly points out the special challenges for dealing with social media texts. It serves as starting point for the technical design of new social media taggers.

The outline of this paper is as follows: Section 2 summarizes the related work and gives an overview of POS tagging. In Section 3, we introduce the new social media text corpus and propose annotation rules for social media texts. Section 4 and Section 5 present our evaluation methodology and corresponding results. Section 6 concludes this work and discusses future research.

2   Related Work

Several papers have been published that deal with automatic POS tagging mainly by following statistical approaches. However, a number of rule-based methods have been proposed in the early stages of POS tagging research. The first rule-based approach has been presented in [9]. One of the latest rule-based methods is proposed in [4]. It yields similar accuracies as statistical approaches. Typical statistical POS taggers make use of two different probabilistic models, a Markov model or a maximum entropy model that captures lexical and contextual information.

Common Markov model taggers are proposed in [13,3]. TreeTagger [13] and TnT [3] are second order Markov models with some smoothing techniques for the estimation of lexical probabilities. TreeTagger utilizes a decision tree for reliable estimation of tag transition probabilities. Maximum entropy based taggers are proposed in [14,5]. These methods use the same baseline maximum entropy model and adapt their approach by using different features in the model. Furthermore, some other machine learning techniques are applied to the problem of automatic POS annotation, e.g., Support Vector Machines [7] and Neural Networks [12].

In [16,6] common POS taggers are evaluated and compared for German. Schneider et al. [16] compare a statistical and a rule-based tagger and point out the performance loss of the rule-based approach applied to unknown words. The performance of five state-of-the-art taggers applied to Web texts is studied in [6]. The corresponding results show that the automatic tagging of Web texts is not yet sufficient and that the accuracy drops significantly for different text genre.

The particular task of tagging non-standardized texts, characterized by frequent unknown words, is addressed in [5,10]. Gadde et al. [5] propose adaptions to the Stanford tagger to handle noisy English text. They evaluate their results based on a Short Message Service (SMS) dataset. They suggest to correct the tags of noisy words in a postprocessing step as well as some preprocessing cleaning techniques to the noise in the given sentence. Gimpel et al. [8] propose a twitter tagger based on a conditional random field (CRF) and adapt their features to twitter characteristics. In [10], the same authors propose some additional word clustering and further improve their method.

3   Corpora

Two corpora are used for training purposes, our social media corpus and a newspaper corpus. First, we introduce *WebCom* a new corpus that contains Web comments collected from *Heise.de*, which is a popular German newsticker site treating different technological topics. The comments for the manual POS annotation are selected from this underlying corpus. In order to obtain a corpus where many kinds of social media characteristics are represented, we select comments from different users. The selection of comments is carried out randomly over different users according to their posting frequencies. Each token is annotated with manually validated POS tags and lemmas. Annotation rules, particularly for social media text characteristics, are given in Section 3.1. A detailed annotation guideline as well as Inter-Annotator Agreement (IAA) studies can be found in [15]. We call the resulting corpus *WebTrain* because it provides supervised data for training purposes. The average POS tag ambiguity of tokens contained in the corpus is 2. This is significantly higher as the ambiguity in German newspaper texts, e.g. 1 for the *TIGER* corpus. Further statistical corpus information is given in Table 1. To the best of our knowledge, *WebTrain* is currently the largest social media text corpus enriched with POS information. However, aiming at training a POS tagger, 36,000 tokens is a relatively small number.

Table 1. Unsupervised and supervised social media text corpora

	WebCom	WebTrain
#Comments:	153,740	429
#Tokens:	15,080,976	36,284
#Words:	360,177	7,830
#Users:	15,007	183

In order to provide a sufficiently large training data amount, we combine *Web-Train* with the *TIGER* treebank [2] newspaper text corpus. It is the largest manually annotated German corpus and contains about 900,000 tokens of German newspaper text, taken from the *Frankfurter Rundschau*. The corpus annotation provides manually validated POS tags, lemmas, morphosyntactic features, and parse trees. For our purposes, only the STTS POS tag information is used.

To have a deeper look in the general applicability of the retrained taggers for social media texts, we create an additional corpus *WebTypes*. It is composed of roughly 5,000 tokens, where comments from different web sites and a corpus extract from the *Dortmunder chat corpus BalaCK 1-b* [1] are annotated in the same way than *WebTrain*. Four different types of social media texts are represented, Merkur newsticker comments, YouTube comments, blog comments, and chat messages.

3.1 Annotation Rules

The STTS tagset was developed 1999 in Stuttgart and has evolved over the years to the standard tagset for the morphosyntactic annotation in German; it provides information about the respective part of speech and its syntactic function. It was developed for the annotation of standardized texts. Until now, no extension for the annotation of the special characteristics of social media texts, e.g., emoticons, is present. Moreover, an extension of the existing tagset is problematic from a technical perspective, since existing NLP methods, e.g., syntactical parsing, require STTS POS tag information. Thus, the existing STTS tagset is used and social media text characteristics are tagged according to their syntactic function in our approach. For instance, emoticons are either at the end of a sentence or at intermediate positions. Therefore, they obtain the tag for sentence final "$." and sentence internal "$(" character. Contrarily, special characters and enumerations are only annotated with the internal character tag. Separated particles of apostrophization, e.g., ([*hab*], *'s - have it*), are tagged for verbs, conjunctions, and interrogative pronouns as *irreflexive personal pronoun* (PPER), *substituting demonstrative pronoun* (PDS), or *article* (ART). Numbers replaced by the corresponding digit in a word are annotated as *attributive adjective* (ADJA) or *proper noun* (NE), depending on the context. The overall annotation rules for particular social media characteristics are given in Table 2. All tags from the first column can be assigned according to the given grammatical context. Exemplary tokens are given in the last column. Note, that the text

Table 2. Annotation rules for social media texts

Tag	Description	Example
$. , $(	Emoticons	:-) , (*_*)
NE	File names, URLs	test.jpg , www.test.de
ITJ , PTKANT	Interaction words, inflectives	lol , seufz , yep
$(	Special characters	#, @, *, i. ii., a) b)
$(, $.	Multiple punctuations	... , !?!
PPER , ART , PDS	Apostrophization	[geht]'s , [wer]'s , [ob]'s
ADJA , NE	Number replacement	10er , 500er

is manually tokenized such that adequate POS annotation can be performed according to the given rules.

4   Evaluation Methodology

For our evaluation, we consider four state-of-the art taggers. We choose the taggers according to their tagging accuracy applied to German standardized texts. Furthermore, all taggers are used in the evaluation section of [6], where the performance of state-of-the art taggers on Web texts is studied. Hence, our results can be compared later to the published results. The selected taggers are the following:

1. TreeTagger [13], a Markov model based tagger using a decision tree for the estimation of tag transition probability.
2. TnT [3], a Markov model tagger that integrates some smoothing techniques for the estimation of lexical probabilities.
3. Stanford [14], a maximum entropy based tagger, integrating different word and tag features.
4. SVMTool [7], a tagger that utilizes support vector machines for classification.

The following evaluations are performed for all taggers, using the proposed default settings for training. Two 10-fold cross validations are carried out to evaluate and compare the tagging accuracy of the four taggers. First, a cross validation on ten equally sized *TIGER* corpus parts is carried out for randomly selected sentences. In the following we call such taggers *TIGER* taggers. Secondly, a 10-fold cross validation is performed on joint-domain training data from the *TIGER* and *WebTrain* corpus. The taggers are trained on a combination of nine *Web-Train* subsets and nine *TIGER* subsets in each validation step. Note, that we use a fix combination of nine *TIGER* subsets here in order to keep a remaining part for testing. We call the resulting ten trained taggers *WebTrain* taggers. For the sake of fair comparison, we use the pre-tokenized data and determine per-word accuracies for the same number of tokens. Furthermore, we study the application of the taggers to social media texts types differing from the training text type. A tagger trained on Web comments from the newsticker site *Heise.de* is, for example tested on blog comments from different blog sites. Based on the cross validation trainings, the performance of all taggers is tested on the *WebType* corpus. Finally, the goal of our work is to analyze for which text characteristics, the tagging accuracies are improved by adding in-domain training data. Therefore, we introduce four categories, motivated by technical challenges of POS tagging, to describe social media text characteristics:

1. *Spoken language character* - The language is borrowed from spoken language and characterized by linguistic irregularities. In German Web comments, verbs are often shortened or merged (e.g., *hab, habs - have, have it*), fill and swear words are used (e.g., *Verdammt - Damn*), reflection periods are verbalized by interjections (e.g., *hmm, äh*) or elliptical constructions are used (e.g., *Entschuldigung! - Sorry!*). Thus, the language is characterized by a lower standardization degree or colloquial style as for newspaper texts.
2. *Dialog form* - A dialogic style characterizes the communication in social media applications. Hence, first and second person singular and plural formulations are predominant. On the other hand, newspaper texts are typically written in third person singular and plural, as this text type has a more descriptive character. Moreover, Web comments are dialogic texts, where many anaphoric expressions (e.g., *die - this, that*) can occur.
3. *Social media language* - The language is characterized by the use of interaction signs such as emoticons (e.g., *:-)*), interaction words (e.g., *lol, rofl*), leetspeak (e.g., *w!k!p3d!4*), word transformations (e.g., *EiPhone*), using mixed languages in the same context and references such as URLs and filenames (e.g., www.google.de).
4. *Informal writing style* - The majority of user posts are written in an informal way. Hence, social media texts suffer from spelling errors, typing errors, abbreviations, missing text and sentence structure (e.g., missing punctuation marks), missing capitalization, character iterations (e.g., *Helloooo*), and multiple punctuation (e.g., *!?!, !!!*).

For a detailed evaluation, the tagging errors are consistently classified into one of the four categories in a manual process by one person. Note, that the resulting classification does not serve for any training but rather for evaluation purposes. Hence, it is just important that the manual classification is performed consistently. However, analyzing the results in that way, is a good starting point for the technical design of new social media text taggers in future work.

5   Evaluation Results

This chapter discusses the results achieved by the previous described evaluation methods. First, we evaluate and compare the performance of the four state-of-the-art taggers applied to social media texts. Results for taggers trained on newspaper *TIGER* texts are compared to those that are additionally trained on *WebTrain*. We particularly point out, that using social media text for training purposes does not negatively effect the tagging accuracies of standardized newspaper texts. Furthermore, trigram statistics show that social media texts differ in their grammatical structure. Finally, in Section 5.2 the application of the *WebTrain* taggers to different social media text types is investigated.

5.1 Tagger Comparison

Comparison starts with a discussion of tagger performance if all test corpora are considered and the taggers are trained on the *TIGER* corpus only. The results are compared to the performance when the taggers are trained on *WebTrain* (added to *TIGER*). More detailed results are presented to investigate the impact of newly learned words and newly seen trigrams.

Results for *TIGER* Training. First we compute the mean tagging accuracies with standard deviations for cross validations performed on *TIGER* and on additional *WebTrain* data. The first row in Table 3 gives the results for the different *TIGER* taggers. Additionally, the first column shows the tagging accuracy achieved by TreeTagger using the standard parameter file (Tree-SPF). Tagging accuracies around 97% are achieved for a comparison to the results from [6]. Slight deviations are observed due to the selection procedure of training sentences. TreeTagger performs worst on *TIGER* data. This is also stated in [6], see Table 3. The second row shows the results achieved by *TIGER* taggers performed on the ten test samples from *WebTrain*. The average tagging accuracies considerably decrease by 8 to 10 percentage points and the standard deviations increase. This can be explained by a different degree of social media text characteristics occurring in the randomly chosen test data.

Results for *WebTrain* Training. We consider the results achieved by *WebTrain* taggers. The second row of Table 4 gives the mean values achieved on *WebTrain* test data. For all state-of-the-art taggers adding in-domain training data leads to an improvement of 5.06 to 5.65 percentage points in average tagging accuracy. Applying TreeTagger results in the maximum average per-word accuracy of 93.72% in contrast to results from Table 3 for standardized *TIGER* texts, where TreeTagger performs the worst. This indicates that the TreeTagger approach is particularly suitable for dealing with the characteristics of non-standardized social media texts. Additionally, we test all *WebTrain* taggers on the held-out TIGER test sample. The corresponding tagging accuracies are depicted in the first row of Table 4. The results demonstrate, that using non-standardized social media texts as additional training data slightly increases the performance for tagging newspaper texts. Note, that the number of tokens for *WebTrain* test and *TIGER* test is the mean value over all cross validation test sets.

Table 3. Tagger evaluation for different text types trained on *TIGER*

Text type	Tree-SPF	TreeTagger	TnT	Stanford	SVM
TIGER test	95.54 ± 0.06	97.18 ± 0.04	97.29 ± 0.05	97.42 ± 0.03	97.45 ± 0.03
WebTrain test	87.08 ± 0.87	88.51 ± 0.99	88.57 ± 1.14	87.74 ± 1.02	87.65 ± 1.13
Merkur comments	94.95	93.11 ± 0.34	90.78 ± 0.73	89.96 ± 0.42	91.64 ± 0.31
Chat messages	81.89	85.63 ± 0.39	84.34 ± 0.24	83.78 ± 0.26	82.80 ± 0.27
YouTube comments	78.88	77.53 ± 0.59	74.85 ± 0.39	74.44 ± 0.55	74.27 ± 0.42
Blog comments	87.98	88.14 ± 0.53	86.93 ± 0.68	86.53 ± 0.51	85.13 ± 0.67

Table 4. Tagger evaluation for different text types trained on *WebTrain*

Text type	#Tokens	TreeTagger	TnT	Stanford	SVM
TIGER test	5,306	97.18 ± 0.03	97.31 ± 0.01	97.44 ± 0.01	97.47 ± 0.01
WebTrain test	3,628	93.72 ± 0.49	93.63 ± 0.37	93.18 ± 0.32	93.30 ± 0.56
Merkur comments	990	94.89 ± 0.38	93.49 ± 0.36	92.46 ± 0.38	93.72 ± 0.41
Chat messages	1,728	89.12 ± 0.18	87.96 ± 0.11	87.81 ± 0.16	86.57 ± 0.13
YouTube comments	1,463	84.03 ± 0.24	81.18 ± 0.19	81.23 ± 0.16	80.56 ± 0.19
Blog comments	815	91.35 ± 0.18	90.46 ± 0.12	90.29 ± 0.17	88.04 ± 0.13

More detailed cross validation results achieved by *WebTrain* taggers are given
in Table 5. Particularly, the accuracy rates are split up into known words and
out-of-vocabulary (unknown) words. We give mean accuracies and standard deviations as well as the percentage of unknown words. In general, unknown words
are such words, that are not known from the training text, i.e., the arising lexicon. Note, that unknown word rates for the same data are partially differing
for different taggers. For instance, TreeTagger excludes cardinal numbers from
unknown word counts. Unknown word rates are roughly 8% for all considered
taggers. This is about 2 percentage points higher than stated for *TIGER* newspaper texts. Particularly, standard deviations for unknown words indicate how
robust a tagger is when social media texts are considered. Stanford tagger is the
most robust tagger. However, the standard deviation of 1.97 is still pretty high.
The highest tagging accuracies of 70.58% for unknown words, with only slightly
higher standard deviations are achieved with the SVMTool. Nevertheless, Tree-
Tagger slightly outperforms the other taggers in total.

Impact of Newly Learned Words. In order to investigate the performance
on social media texts we carry out different training approaches. Exemplary experiments are presented for the TreeTagger. Table 6 shows the results achieved
by using differently trained taggers, i.e., TreeTagger with the standard parameter file, *TIGER* tagger, *TIGER* tagger with an auxiliary lexicon created from
WebTrain, and *WebTrain* tagger. The auxiliary lexicon covers the words/tokens
that are contained in *WebTrain* texts including their corresponding set of possible tags. Neither lexical probabilities for such words nor trigrams of such texts
are given. Tagging is performed on *WebTrain* test data. The results achieved by

Table 5. Results for 10-fold cross validation trained joint-domain data using *WebTrain*

	TreeTagger	TnT	Stanford	SVM
Total	93.72 ± 0.49	93.63 ± 0.37	93.18 ± 0.32	93.30 ± 0.56
Known	95.83 ± 0.43	95.81 ± 0.51	95.61 ± 0.40	95.58 ± 0.45
Unknown	67.98 ± 3.14	70.58 ± 2.08	68.14 ± 1.97	69.33 ± 2.54
Percentage unknowns	7.58 ± 0.75	8.65 ± 0.62	8.81 ± 0.62	8.65 ± 0.62

Table 6. Detailed TreeTagger tagging accuracies for different training approaches

	Tree-SPF	*TIGER*	*TIGER* + Web Lex.	*WebTrain*
Total	87.08 ± 0.87	88.51 ± 0.99	92.63 ± 0.68	93.72 ± 0.49
Known	92.05 ± 0.53	94.47 ± 0.57	94.74 ± 0.53	95.83 ± 0.43
Unknown	44.77 ± 2.46	54.13 ± 3.15	66.47 ± 3.04	67.98 ± 3.14
Percentage unknown	10.50 ± 0.76	14.71 ± 0.96	7.58 ± 0.75	7.58 ± 0.75

using the standard parameter file for unknown words show a high performance loss. This can be explained by the fact that for unknown words the so called open-class tags are restricted to 7 STTS tags for nouns, named entities, verbs, and adjectives. This restriction is inappropriate for the task of tagging social media texts, where unknown words can almost be of any word class. The number of unknown words is reduced by one half when comparing *TIGER* tagger results with *WebTrain* tagger results. Moreover, the tagging accuracy for the lower number of unknowns is increased by more than 10 percentage points.

Impact of Newly Seen Trigrams. We use the *WebTrain* text corpus for training instead of only considering it as a word lexicon. This leads to further tagging improvement by more than one percentage point, see the right column of Table 6. This can be explained by different grammatical structures of social media texts, which need to be learned from a sufficient amount of in-domain training data. Finally, we demonstrate how different the grammatical structure in social media texts is compared to newspaper texts. STTS tag trigram frequencies are calculated for both corpora *TIGER* and *WebTrain*. The overall results are depicted in Table 7. The third column shows the ratio between different trigrams and their frequencies for the different corpora. Results illustrate the higher variability in social media texts, which is ten times higher than in newspaper texts. Particularly, we compare statistics for tag trigrams that occur in *WebTrain* texts but are unknown from the *TIGER* corpus. The statistics are given in the last row. *WebTrain* texts contain 18% new trigrams, that never occur in the newspaper corpus *TIGER*. Those trigrams constitute 6% frequency of all *WebTrain* trigram counts. Particularly, for those trigrams the ratio/variability is increasing by a factor of three. Both results motivate the need of in-domain training data for reliable estimation of transition probabilities, e.g., for trigrams.

Table 7. Trigram comparison for *TIGER* and *WebTrain* corpora

	Trigrams	Trigram frequencies	Ratio
Total *WebTrain*	7,215	36,282	0.20
Total *TIGER*	16,563	888,982	0.02
Only in *WebTrain*	1,290	2,120	0.61

5.2 Results for Different Social Media Text Types

In this section, we study the application of taggers to different social media text types, where the taggers are not trained for the particular type. To illustrate the improvements, Table 3 shows tagging accuracies and standard deviations for all taggers trained on *TIGER*. Table 4 depicts the improved tagging accuracies that are achieved by *WebTrain* taggers. Application of joint-domain trained taggers leads to a consistent performance increase between approximately 2 and 7 percentage points for different social media text types. Considerable improvements can be observed for chat and YouTube data, which are highly characterized by a dialogue form. Moreover, TreeTagger outperforms TnT, Stanford, and SVMTool for all considered social media text types.

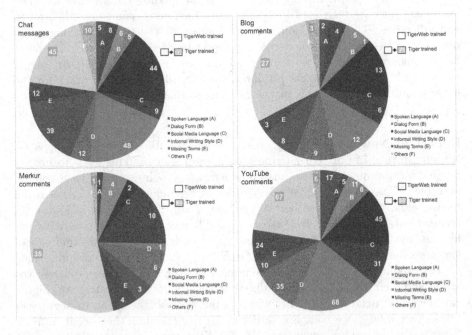

Fig. 1. Error classification and improvement for different social media text types

Finally, we evaluate the results for all social media text types with respect to the four different characterization categories introduced in Section 4. Therefore, we filter and classify all words which are not correctly tagged by using

TreeTagger trained on *TIGER* and on joint-domain training *TIGER* and *Web-Train*, respectively. Note, that the social media text categories are complemented by two more categories, *missing terms* and *others*. The category *missing terms* covers topic specific nouns and named entities. The category *others* comprises all other occurring wrongly tagged words (e.g. *aber - but, die - the*), which are not related to any of these categories and can also occur in standardized texts. Figure 1 depicts absolute errors for each category. The shaded areas illustrate the absolute error reduction for each particular category. The corresponding total number of tokens for each type are depicted in Table 4. Applying the *WebTrain* TreeTagger, the errors made for all four social media categories by the *TIGER* TreeTagger can be reduced from 26% to 71% . Errors are reduced up to 86% for the *social media language* category. Using in-domain training data, effectively reduces *missing terms* errors by more than a third for all text types. The error rates for the category *others* can hardly be improved. The enrichment with in-domain training enables the special handling of social media text characteristics, particularly for *social media language* and *informal writing style* categories. In total, a significant error reduction can be achieved over all categories.

6   Conclusions

We have shown that the performance of state-of-the-art POS taggers can significantly be improved for social media texts. The improvement is achieved by taking training data from social media texts into account. We have created a new social media text corpus *WebTrain* that contains 38,000 manually annotated tokens. It can be used to retrain such taggers. We introduce an adequate STTS annotation guideline for social media texts. To fulfill the requirement of other NLP methods, we use the original STTS tag set without any extensions.

For all state-of-the-art taggers, adding in-domain training data leads to a significant improvement of more than five percentage points for the tagging accuracy. TreeTagger cross validation leads to a maximum average per-word accuracy of 93.72%. Moreover, TreeTagger outperforms TnT, Stanford, and SVMTool for all considered social media text types. Taggers trained on Web comments can successfully be used for different text types. Applying the joint-domain trained taggers leads to a consistent performance increase between approximately 2 and 7 percentage points for different text types. Considerable improvements are obtained for chat and YouTube data, which are highly characterized by a dialogue form. However, the overall accuracies of 89% and 84% demand for further investigations. Enrichment of the newspaper corpus by social media text data, leads to a slightly improved evaluation on newspaper texts. Hence, the enhanced training data improves tagging results on all kind of investigated text types.

Finally, we have shown that the grammatical structure in social media texts differs. The new grammatical structures are learned from a sufficient amount of in-domain training data and account for a considerable improvement of the tagging accuracy.

Beyond providing in-domain training data for the enhancement of POS tagging accuracy for social media texts, we currently work on adaptions to existing tagger models. We particularly focus on the parameter estimation for unknown words to further improve accuracies, particularly for chat and YouTube data.

References

1. Beißwenger, M.: Corpora zur computervermittelten (internetbasierten) Kommunikation. Zeitschrift für Germanistische Linguistik 35, 496–503 (2007)
2. Brants, S., Dipper, S., Eisenberg, P., Hansen-Schirra, S., König, E., Lezius, W., Rohrer, C., Smith, G., Uszkoreit, H.: TIGER: Linguistic Interpretation of a German Corpus. In: Research on Language & Computation, pp. 597–620 (2004)
3. Brants, T.: TnT – A Statistical Part-of-Speech Tagger. In: Proceedings of the 6th Applied Natural Language Processing Conference, pp. 224–231 (2000)
4. Brill, E.: A Simple Rule-based Part of Speech Tagger. In: Proceedings of the Third Conference on Applied Natural Language Processing, pp. 152–155 (1992)
5. Gadde, P., Subramaniam, L.V., Faruquie, T.A.: Adapting a WSJ Trained Part-of-Speech Tagger to Noisy Text: Preliminary Results. In: Proceedings of the 2011 Joint Workshop on Multilingual OCR and Analytics for Noisy Unstructured Text Data, pp. 5:1–5:8 (2011)
6. Giesbrecht, E., Evert, S.: Is Part-of-Speech Tagging a Solved Task? An Evaluation of POS Taggers for the German Web as Corpus. In: Proceedings of the Fifth Web as Corpus Workshop, pp. 27–35 (2009)
7. Giménez, J., Màrquez, L.: Svmtool: A General POS Tagger Generator Based on Support Vector Machines. In: Proceedings of the 4th International Conference on Language Resources and Evaluation, pp. 43–46 (2004)
8. Gimpel, K., Schneider, N., O'Connor, B., Das, D., Mills, D., Eisenstein, J., Heilman, M., Yogatama, D., Flanigan, J., Smith, N.A.: Part-of-speech tagging for Twitter: annotation, features, and experiments. In: Proceedings of the 49th Annual Meeting of the Association for Computational Linguistics, pp. 42–47 (2011)
9. Klein, S., Simmons, R.F.: A Computational Approach to Grammatical Coding of English Words. J. ACM 10, 334–347 (1963)
10. Owoputi, O., O'Connor, B., Dyer, C., Gimpel, K., Schneider, N.: Part-of-Speech Tagging for Twitter: Word Clusters and Other Advances. Technical report, School of Computer Science, Carnegie Mellon University (2012)
11. Schiller, A., Teufel, S., Stöckert, C., Thielen, C.: Guidelines für das Tagging deutscher Textcorpora mit STTS. University of Stuttgart (1999)
12. Schmid, H.: Part-of-Speech Tagging With Neural Networks. In: Proceedings of the 15th Conference on Computational Linguistics, pp. 172–176 (1994)
13. Schmid, H.: Improvements in Part-of-Speech Tagging With an Application to German. In: Proceedings of the ACL SIGDAT-Workshop, pp. 47–50 (1995)
14. Toutanova, K., Klein, D., Manning, C.D., Singer, Y.: Feature-rich Part-of-Speech Tagging With a Cyclic Dependency Network. In: Proceedings of Human Language Technology Conference, pp. 173–180 (2003)
15. Trevisan, B., Neunerdt, M., Jakobs, E.-M.: A multi-level annotation model for fine-grained opinion detection in German blog comments. In: Proceedings of KONVENS 2012, pp. 179–188 (2012)
16. Volk, M., Schneider, G.: Comparing a statistical and a rule-based tagger for German. In: Proceedings of the 4th Conference on Natural Language Processing, x pp. 125–137 (1998)

Summarizing Answers
for Community Question Answer Services

Vinay Pande[1], Tanmoy Mukherjee[1], and Vasudeva Varma[2]

[1] International Institute of Information Technology, Hyderabad
{vinay.pande,tanmoy.mukherjee}@research.iiit.ac.in
[2] International Institute of Information Technology, Hyderabad
vv@iiit.ac.in

Abstract. This paper presents a novel answer summarization approach for community Question Answering services (cQAs) to address the problem of "incomplete answer", i.e., missing valuable information from the "best answer" of a complex multi-sentence question, which can be obtained from other answers to the same question. Our method automatically generate a novel and non-redundant summary from cQA answers using structured determinantal point processes (SDPP). Experimental evaluation on sample dataset from Yahoo Answers shows significant improvement over baseline approaches.

1 Introduction

Community Question and Answering services (cQAs) like Yahoo Answers[1], Stackoverflow[2] allow users to post questions of their interests and other users share their knowledge by providing answers to the questions. A question often has multiple answers and a best answer is selected based on criteria for the portal. This {question, best answer} pair is then stored and indexed for future re-use such as question retrieval. cQA are becoming quite popular for their ability to provide precise and concise answers. Cong et al.(2008) noticed that 90% of forums contain question-answer knowledge. The knowledge contained in these question-answer pairs could be a huge source of information for search which often have natural language questions. In general, this system works good for factoid QA setting, where answer is often points to single a named entity like person, location or time. However for more sophisticated multi-sentence questions, the best answer chosen might be "incomplete" (Chan et al. 2012) as such questions have multiple sub-questions in different contexts and user may be interested in all of them. In this case, the "best" answer covering few "aspects" may not be a good or ideal choice. In early literature, Liu et al.(2008) noticed that no more than 48% of the best 400 answers in 4 popular Yahoo! answer categories were unique best answer. Table 1 shows example of question with incomplete answer problem. Here, asker want to know, "why image size is small for Instagram in Android?". But "Best Answer" is not providing any information regarding that and hence is not of any future re-use. However, some

---

[1] http://answers.yahoo.com/
[2] http://stackoverflow.com/

I. Gurevych, C. Biemann, and T. Zesch (Eds.): GSCL 2013, LNAI 8105, pp. 151–161, 2013.
© Springer-Verlag Berlin Heidelberg 2013

answers given by other users contain the required information. Hence, summarizing this answers will definitely help users to find proper answers to their questions.

In general, users are more interested in multi-sentence questions having long answers containing more information. In fact, it is often the case, that a complex multi-sentence question could be answered from multiple aspects by different people focusing on different sub questions. Since "everyone knows something" (Adamic et al. 2008), the use of a single best answer often misses valuable human generated information contained in other answers. Therefore, in this work, we addressed "incomplete answer" problem by summarizing user answers using graph based approach. We try to find out multiple threads containing diverse and informative sentences from the answers using structured determinantal point processes (SDPP), we then choose the best thread as a summary using weight function on nodes and edges.

Table 1. An example of question with incomplete answer problem from Yahoo! Answers. The best answer seems to miss valuable information and hence will not be ideal for re-use when similar question is asked again.

Question
Why are all my images for Instagram for Android too small? I have the sidekick 4g and every time I try to upload to Instagram, an error appears saying "photo is too small". My camera resolution has been at the highest ever since I got the phone so its not that. Any idea why this is happening?
Best Answer
I'm having the same problem. I have a Galaxy S. It sucks because I've waited this entire time for instagram and now its not working. I hope somebody comes up with a working solution soon
Sample of Other Answers
I have the sidekick 4g and i went to my camera -click the little area at the side of screen -Then click on settings in the top right corner -click the arrow down to page 2 -by the resolution click it and select 2048x1232 this will work. the 1600x960 might not work for instagram so to be on the safe side put the one that is in the instructions above. Hope this helped you :)
Launch Instagram. Click on the icon that looks like an ID card on the bottom right. Then at the top right hand corner click the icon that has 3 small squares in a row. Scroll down and click on "Camera Settings". Select, "Use Instagram's Advanced Camera". Problem fixed.

We conduct experiments on a Yahoo! Answers dataset. The experimental results show that the proposed model improve the performance significantly(in terms of precision, recall and F1 measures) as well as the ROUGE-1, ROUGE-2 and ROUGE-L measures as compared to the state-of-the-art methods, such as Support Vector Machines (SVM), Logistic Regression (LR) and gCRF (Chan et al. 2012).

Remaining paper is organized as follows: Section 2 gives definitions and related work. Section 3 explains our approach and Detrimental Point Process. Section 4 contains details about experiments and Section 5 concludes the approach.

2 Definitions and Related Work

2.1 Complex Multi-sentence Question

In cQA scenario, a question often contains main title with one or more sub-questions and a short description given by the asker. We treat the main questions and contexts from the description as single complex multi-sentence question.

2.2 Incomplete Answer Problem

A given "best answer" is incomplete answer if it is voted below certain star ratings or average similarity between best answer and all sub-questions is below some threshold (Chan et al. 2012)

2.3 Related Work

Many people attempted the problem of answer completeness in cQA. Wang et al.(2010) tried to solve the incomplete answer problem by segmenting main-question into several sub-questions with their contexts. They then sequentially retrieve best answers for every question similar to sub-question. However due to errors in segmentation, this strategy may retrieve best answers of the questions in totally different context and could not combine multiple independent best answers of sub questions seamlessly and may introduce redundancy in final answer.

Liu et al.(2008) applied clustering algorithms for answer summarization. They manually classified both questions and answers into different taxonomies by utilizing textual features for open and opinion type questions

Tomasoni and Huang(2010) introduced four characteristics of summarized answer and combined them in an additional model as well as a multiplicative model using metadata.

Yang et al.(2011) enhanced the performance of social document summarization with user generated content like tweets by employing a dual wing factor graph

Wang et al.(2011) used general CRFs to learn online discussion structures such as the replying relationship and presented a detailed description of their feature designs for sites and edges embedded in discussion thread structures.

Chan et al.(2012) tackle the answer summary task as a sequential labeling process under the general Conditional Random Fields (gCRF) framework. They incorporated four different contextual factors based on question segmentation to model the local and non-local semantic sentence interactions to address the problem of redundancy and information novelty. They also exploited various textual and non-textual question answering features and proposed a group L_1 regularization approach in the CRF model for automatic optimal feature learning.

In previous work on SDPPs, Kulesza and Taskar(2010) derived exact polynomial-time algorithms for sampling and other inference. However, their experiments involved feature vectors of only 32 dimensions.

Gillenwater et al.(2012) solved a problem of finding set of diverse and salient threads from document collections using SDPPs. They also proposed a method to determine importance of document threads using weight functions on nodes and edges of the graph.

In this paper we use SDPPs to create summary of answers in cQA by finding diverse set of informative threads using different textual and non-textual features.

3   Our Approach

We assume that all the answers in the question answering thread can be represented as a connected graph G = (V,E). Where, every sentence in answers is a node and edge indicates similarity between two nodes defined by some similarity metric. Our goal is to find a path containing nodes that are individually coherent and together cover the most important information from the graph as shown in Figure 1 (Gillenwater et al. 2012). We also assume there is a weight function w defined on nodes and edges of graph,

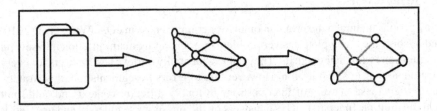

Fig. 1. We first build a graph from the sentences, using measures of importance and relatedness to weight nodes (sentences) and build edges (similarity between sentences). Then, from this graph, we extract a diverse, salient set of threads to summarize the answers.

which measures the importance of sentences and the relative strength of the similarity between them, we can formally define the weight of a path (or thread) $y = (y^{(1)}, y^{(2)}, ..., y^{(T)})$, $(y^{(t)}, y^{(t+1)}) \in E$ as (Gillenwater et al. 2012) :

$$w(y) = \sum_{t=1}^{T} w(y^{(t)}) + \sum_{t=1}^{T} w(y^{(t)}, y^{t+1}) \tag{1}$$

Given this framework, our goal is to develop a probabilistic model over sets of k threads of length T , favoring sets whose threads have large weight but are also distinct from one another. In other words, a high-probability set under the model should include sentences that are both important and diverse.

We solve this problem using Structured Determinantal point processes.

3.1   Structured Determinantal Point Processes

A DPP is a type of distribution over subsets. Formally, a DPP P on a set of items $\mathcal{Y} = \{y_1, ..., y_N\}$ is a probability measure on $2^{\mathcal{Y}}$, the set of all subsets of $\mathcal{Y}$. For every $Y \subseteq \mathcal{Y}$ we have:

$$P(Y) = \frac{det(L_Y)}{\sum_{Y \subseteq \mathcal{Y}} det(L_Y)} = \frac{det(L_Y)}{det(L + I)} \tag{2}$$

where, I is the N X N identity matrix and L is a positive semi-definite matrix given by:

$$L_{ij} = q(y_i)\phi(y_i)^T \phi(y_j)q(y_j) \tag{3}$$

where we can think of $q(y_i) \in \mathbb{R}^+$ as the quality of an item y_i , and $\phi(y_i) \in \mathbb{R}^D, \|\phi(y_i)\|_2 = 1$ is a normalized D-dimensional feature vector over nodes such that $\phi(y_i)^T \phi(y_j) \in [-1, 1]$ is a measure of similarity between items y_i and y_j . This simple definition gives rise to a distribution that places most of its weight on sets that are both high quality and diverse.

In order to allow for efficient normalization and sampling, Kulesza and Taskar(2010) introduced structured DPPs (SDPPs) to efficiently handle $\mathcal{Y}$ containing exponentially many structures. SDPPs assume a factorization of the quality score $q(y_i)$ and similarity score $\phi(y_i)^T \phi(y_j)$ into parts, decomposing quality multiplicatively and similarity additively:

$$q(y_i) = \prod_{t=1}^{T} q(y_i^{(t)}), \phi(y_i) = \sum_{t=1}^{T} \phi(y_i^{(t)}) \tag{4}$$

Thus using this model, we will get set of diverse threads spanning across the graph, where weight of each thread can be calculated using equation 1. The thread with maximum weight contains diverse and most informative summary of the answers. Determinants are closely related to volumes; in particular, $det(L_Y)$ is proportional to the volume spanned by the vectors $q(y_i)\phi(y_i)$ for $y_i \in Y$. Thus, sets with high-quality, diverse items have the highest probability.

The sampling algorithm requires $O(Tn^2D^2)$ time, where T is the number of threads, n is number of nodes and D is numbed of features. Detailed explanation of this model is given in Kulesza and Taskar Kulesza and Taskar 2010.

After discussing our approach, we will discuss graph creation process in next section.

4 Creating Graph

For creating Graph, We extract sentences from each answer using nltk[3] toolkit. While doing so we capture following features for each sentence:

4.1 Sentence Level Features

Sentence Length: Long sentence may contains more information and hence add more information to summary. We use sentence length as a feature after removing stopwords from the sentence.

Position in Answer: Generally sentences at the starting or at the end of the answer are more important for summary as they contains more information. We use binary features to check if the given sentence is present at the starting of answer(first 3 sentences), or at the end of answer (last 3 sentences), or at the middle of the answer.

Link: A link in sentence may point to some important resource regarding the context of question. Hence, sentence containing links are more important while generating summary. We use binary feature to check if sentence contains some link.

Number of Stopwords: A sentence containing more stopwords may be a spam sentence.

Number of Capital Words: Capital words are mostly named entities related to question and hence should be added to summary. We use binary feature to check if sentence has capital words in it.

Similarity to Sub-question: We calculate the similarity of sentence to each sub-question and use it as a feature. Sentence having more similarity to the question can be more informative while creating summary.

4.2 Answer Level Features

Answer level features can be useful while calculating node quality.

User Ratings: Number of up-votes and down-votes given to the answer by asker or other users.

Answer Length: Long answer contains detailed discussion about various sub-questions and hence important for summary.

Similarity to sub-question: We use average similarity of the answer to all the sub-questions as a feature while calculating node importance.

4.3 Similarity

We use two metrices for calculating word level and semantic level similarity between sentences-sentence and sentence-subquestion pair.

---

[3] http://nltk.org/

Normalized Cosine Similarity: We calculate cosine similarity between sentences i and j using formula:

$$S_{ij} = \frac{\sum_{w \in W} Count_i(w) Count_j(w)}{\sqrt{\sum_{w \in W} Count_i(w)} \sqrt{\sum_{w \in W} Count_j(w)}} \tag{5}$$

Where, W is a dictionary of all the words and $Count_i(x)$ is a count of word x in sentence i.

Semantic Similarity: We calculated semantic similarity between two sentences using Wordnet. For calculating semantic similarity between two sentence x and y we compare all pairs of synsets present in the sentences using formula:

$$M(x, y) = 2X \sum_{(w_1, w_2) \in M(x, y)} \frac{sim(w_1, w_2)}{|x||y|} \tag{6}$$

Where, Similarity between synset w_1 and w_2 is calculated using length of the path between them in wordnet.

For a given post, number of answers are limited and hence total number of sentences per post is also less. In our test data set, average number of answer sentences per post is 80. Hence, we keep the edges between all the nodes while calculating the threads without pruning the edged for sparse matrix.

5   Experiments and Results

5.1   Dataset

We performed our experiments on Yahoo Answers! and Stackoverflow datasets. Our original dataset contains 5000 questions from Yahoo Answers and 3000 questions from Stackoverflow. Before performing experiments we filtered out questions having less than 5 answers so that we will get sufficient data for summarization per question.

We then extract incomplete answer questions from the dataset. We treat all the questions as incomplete answer question if
1) average similarity between best answer and all subquestions is less than 0.6 or
2) User has given less than 4 stars to the best answer.

After pre-processing, we randomly select 500 questions each from Yahoo Answers and stackoverflow datasets as our sample dataset. We perform all our operations on this sample dataset only.

For question segmentation we use 2 step segmentation method
(Chan et al. 2012,Ding et al. 2008,Wang et al. 2010) as follows:

Step 1: Classify every sentence in the multi-question context (title + description) into question sentences and non-question sentences. For classification we use question mark and 5W1H features.

Step 2: Assign every contextual sentence to some question sentence using semantic similarity.

5.2 SDPP

We use matlab implementation of SDPP available at `www.eecs.umich.edu/~kulesza/code/dpp.tgz` on the graph created using step explained in the previous section. SDDPs needs various factors for calculating set of diverse threads. We calculate values of these factors as follows:

1. **Calculating Node Quality:** We used 4 features to calculate node importance.

   1. User ratings(U_i) – normalized to [0,1] using maximum ratings value
   2. Position (P_i) – Importance of sentence is inversely proportional to position in answer
   3. Sentence length(SL_i) – Importance is proportional to length of sentences
   4. Similarity to question(SQ_i) – Average similarity to all subquestions calculated using similarity metric

We calculate the Node quality using the formula:

$$q(y_i) = \lambda_1 U_i + \lambda_2 1/P_i + \lambda_3 SL_i + \lambda_4 SQ_i \tag{7}$$

2. **Calculating Weight between Edges**
We calculated the weight between edges by calculating similarity between two sentences using features as explained in previous section.

3. **Similarity Feature Map ϕ**
Finally, we build a similarity feature map ϕ to encourage diversity. We represent each sentence by top 10 sentences to which it is most similar according to similarity metric. This gives us similarity feature map for each node containing exactly 10 non-zero values. The dot product between the similarity features of two sentences is thus proportional to the fraction of top-10 similar sentences they have in common.

5.3 Evaluation Measures

By considering summarization as a bi-classification problem, where each sentence is classified into summary sentence or non-summary sentence, we can use precision, recall and F1 measure for evaluating classification accuracy (Shen et al. 2007). We measure precision, recall and F1 measure using ROUGE-1, ROUGE-2 and ROUGE-L measures (Lin 2004).

5.4 Results

We use Support Vector Machines(SVM) and Logistic Regression(LR) as a baseline for classification problem, both considered as state of the art systems for classification. For comparing summarization performance we used Linear CRF (LCRF) as a baseline, which summarize single document text (Shen et al. 2007). We also compare our results with global CRF(gCRF) approach, which summarizes answers for multi-sentence questions in cQA services (Chan et al. 2012)

We evaluate performance of our approach for two setups - 1) Graph created using Cosine Similarity and 2) Graph created using semantic similarity.

Graph created using cosine similarity deals with sentence level similarity and hence prone to semantic level redundancy. That means same thing expressed in different words is difficult to catch using cosine similarity. Semantic similarity deals with more deeper meaning of sentences and hence connect sentences with same meaning more effectively.

We perform all the experiments using 10-fold cross validation. We divide our dataset into 10 parts (each part contains 100 questions and their answers). We then run the experiments on all the groups. Final results are average of results obtained from all the runs. For ROUGE evaluation, we invite group of graduate candidate students to write manual summaries. For each question-answer post, we provide a set of sub-questions to the annotator and asked them to write summary containing answers to as many sub-questions as possible using sentences in the answer threads from the given post. The manually generated summary may contains sentence level redundancy but it helps us to check coverage of our approach.

Table 2. shows comparison of our approach with baseline methods using Yahoo Answers dataset.

Table 2. Precision, Recall and F1 values of ROUGE-1, ROUGE-2 and ROUGE-3 for baselines SVM, LR, LCRF and gCRF and for our approach using cosine similarity(CS) and semantic similarity(SS) on Yahoo Answers dataset

Model	R1-P	R1-R	R1-F1	R2-P	R2-R	R2-F1	RL-P	RL-R	RL-F1
SVM	79.1%	52.3%	62.9%	71.7%	41.4%	52.2%	67.1%	36.4%	47.5%
LR	75.2%	57.5%	65.1%	66.1%	48.5%	56.1%	61.4%	43.3%	50.8%
LCRF	78.7%	61.8%	69.3%	71.4%	54.1%	61.6%	67.1%	49.6%	57.0%
gCRF	86.5%	68.3%	76.4%	82.6%	61.5%	70.5%	80.4%	58.2%	67.5%
Our Approach(CS)	85.3%	69.2%	76.4%	81.7%	62.3%	70.6%	79.1%	59.3%	67.7%
Our Approach(SS)	88.2%	70.1%	78.1%	83.8%	62.8%	71.8%	82.1%	60.1%	69.3%

From Table 2, our approach with Semantic Similarity (SS) improves the performance in terms of precision, recall and F1 score on all three measurements of ROUGE-1, ROUGE-2 and ROUGE-L by a significant margin compared to other baselines due to the use of local and non-local contextual factors while calculating SDDPs. we observe that our question segmentation method can enhance the recall of the summaries significantly due to the more fine-grained modeling of sub questions. ROUGE offer better measurements in modeling user needs as they care more about the recall and precision of N-grams as well as common substrings to the reference summary rather than the whole sentence. Hence, improvement in ROUGE measures are more encouraging than those of the average classification accuracy for answer summarization. In general, the experimental results show that our proposed method is more effective than other baselines in answer summarization for addressing the incomplete answer problem in cQAs.

5.5 Sample Summary

Table 3 shows the generated summary of the example question which is previously illustrated in Table 1 in the introduction section.

Table 3. Sample summary generated for question in Table 1 using our approach

Question
Why are all my images for Instagram for Android too small? I have the sidekick 4g and every time I try to upload to Instagram, an error appears saying "photo is too small". My camera resolution has been at the highest ever since I got the phone so its not that. Any idea why this is happening?
Best Answer
I'm having the same problem. I have a Galaxy S. It sucks because I've waited this entire time for instagram and now its not working. I hope somebody comes up with a working solution soon
Summary using our approach
I have the sidekick 4g and i went to my camera -click the little area at the side of screen -Then click on settings in the top right corner -click the arrow down to page 2 -by the resolution click it and select 2048x1232 this will work. Launch Instagram. Select, "Use Instagram's Advanced Camera".

The best answer available in the system and the summarized answer generated by our model are compared in Table 3. The summarized answer contains more valuable information about the original multisentence question, as it gives multiple solutions to change resolution of images in Instagram using Android. Storing and indexing this summarized answer in question archives should provide a better choice for answer reuse in question retrieval of cQAs.

5.6 Conclusions

We present a novel way to answer "incomplete answer problem" by creating summary of all the answers for a question in the cQA servers. Our main contribution are that we proposed a systematic way to create summary from the sentences by creating graphs using different features. Our method show significant improvement over other methods of answer summarization.

References

Adamic, L.A., Zhang, J., Bakshy, E., Ackerman, M.S.: Knowledge sharing and yahoo answers: everyone knows something. In: Proceedings of WWW (2008)

Chan, W., Zhou, X., Wang, W., Chua, T.-S.: Community answer summarization for multi-sentence question with group 11 regularization. In: Proceedings of the 50th Annual Meeting of the Association for Computational Linguistics: Long Papers. Association for Computational Linguistics, vol. 1, pp. 582–591 (2012)

Cong, G., Wang, L., Lin, C.-Y., Song, Y.-I., Sun, Y.: Finding question-answer pairs from online forums. In: Proceedings of the 31st Annual International ACM SIGIR Conference on Research and Development in Information Retrieval, pp. 467–474. ACM (2008)

Ding, S., Cong, G., Lin, C.Y., Zhu, X.: Using conditional random fields to extract contexts and answers of questions from online forums. In: Proceedings of ACL 2008: HLT (2008)

Gillenwater, J., Kulesza, A., Taskar, B.: Discovering diverse and salient threads in document collections. In: Proceedings of the 2012 Joint Conference on Empirical Methods in Natural Language Processing and Computational Natural Language Learning. Association for Computational Linguistics, pp. 710–720 (2012)

Kulesza, A., Taskar, B.: Structured determinantal point processes (2010)

Lin, C.-Y.: Rouge: A package for automatic evaluation of summaries. In: Text Summarization Branches Out: Proceedings of the ACL 2004 Workshop, pp. 74–81 (2004)

Liu, Y., Bian, J., Agichtein, E.: Predicting information seeker satisfaction in community question answering. In: Proceedings of the 31st Annual International ACM SIGIR Conference on Research and Development in Information Retrieval, pp. 483–490. ACM (2008)

Shen, D., Sun, J.-T., Li, H., Yang, Q., Chen, Z.: Document summarization using conditional random fields. In: Proceedings of the 20th International Joint Conference on Artifical Intelligence, vol. 7, pp. 2862–2867 (2007)

Tomasoni, M., Huang, M.: Metadata-aware measures for answer summarization in community question answering. In: Proceedings of the 48th Annual Meeting of the Association for Computational Linguistics. Association for Computational Linguistics, pp. 760–769 (2010)

Wang, K., Ming, Z.-Y., Hu, X., Chua, T.-S.: Segmentation of multi-sentence questions: towards effective question retrieval in cqa services. In: Proceedings of the 33rd International ACM SIGIR Conference on Research and Development in Information Retrieval, pp. 387–394. ACM (2010)

Wang, H., Wang, C., Zhai, C., Han, J.: Learning online discussion structures by conditional random fields. In: Proceedings of the 34th International ACM SIGIR Conference on Research and Development in Information Retrieval, pp. 435–444. ACM (2011)

Yang, Z., Cai, K., Tang, J., Zhang, L., Su, Z., Li, J.: Social context summarization. In: Proceedings of the 34th International ACM SIGIR Conference on Research and Development in Information Retrieval, pp. 255–264. ACM (2011)

Fine-Grained POS Tagging of German Tweets

Ines Rehbein*

SFB 632 "Information Structure"
German Departement
Potsdam University, Germany
irehbein@uni-potsdam.de

Abstract. This paper presents the first work on POS tagging German Twitter data, showing that despite the noisy and often cryptic nature of the data a fine-grained analysis of POS tags on Twitter microtext is feasible. Our CRF-based tagger achieves an accuracy of around 89% when trained on LDA word clusters, features from an automatically created dictionary and additional out-of-domain training data.

Keywords: POS tagging, Twitter, user-generated content.

1 Introduction

These days, part-of-speech (POS) tagging of canonical, written English seems like a solved problem with accuracies around 98%. However, when applying the same tagger to text from other domains such as web data or spoken language, the tagging accuracy decreases dramatically, and even more so for less resourced languages or languages with a richer morphology than English. A case in point is German where the state-of-the-art for POS tagging newspaper text is in the same range as the one for English, while tagging accuracies for out-of-domain data like Twitter show a substantial decrease for both languages [1, 2].

For English, a POS tagger for Twitter data already exists [3, 4] which provides coarse-grained analyses for English microtext with an accuracy around 92% [4]. Remarkably, these results are in the same range as the results reported for inter-annotator agreement of human annotators on POS tagging tweets [3]. This raises the question whether we have reached an upper bound caused by the often cryptic nature of the data which makes complete disambiguation impossible, or whether there is still room for improvement.

In the paper, we follow up on this question and, in contrast to [3, 4], target a *fine-grained* POS analysis of German data from social media, facing a situation where only a small amount of annotated training data is available and where,

* This work was supported by a grant from the German Research Association (DFG) awarded to the Collaborative Research Centre (SFB) 632 "Information Structure". I gratefully acknowledge the work of our annotator, Emiel Visser. I'd also like to thank the anonymous reviewers for their helpful comments.

I. Gurevych, C. Biemann, and T. Zesch (Eds.): GSCL 2013, LNAI 8105, pp. 162–175, 2013.

in contrast to English, we do expect the richer morphology to aggravate the data sparseness problem. We develop a fine-grained POS tagger for German tweets based on Conditional Random Fields (CRF) and show that after applying different techniques of domain adaptation we are able to obtain an accuracy close to 90%.

The paper is structured as follows. Section 2 reviews related work on POS tagging tweets and Section 3 presents the data and annotation scheme used in our experiments. In Section 4 we describe our experimental setup and the features used in the experiments. We explain our word clustering approach using Latent Dirichlet Allocation (LDA) and the use of additional out-of-domain training data to increase tagging accuracy. In Section 5 we report results for the different techniques. Section 6 presents an error analysis, and we conclude and outline future work in Section 7.

2   Related Work

Some work has been done on POS tagging English tweets. Foster et al. [2] annotate a small treebank of 519 sentences from Twitter, using the Penn Treebank (PTB) annotation scheme. They report a POS tagging accuracy of 84,1% for an SVM-based tagger. The same tagger achieved 96,3% on data from the PTB.

Ritter et al. [5] adapt a CRF-based tagger to Twitter data and present a tagging accuracy of 88,3% using the full 45 tags from the PTB and 4 additional (unambiguous) tags for twitter-specific phenomena (retweets, at-mentions, hashtags and urls; see Section 3.1 for details on the definition of these tags).

Gimpel et al. [3] and Owoputi et al. [4] developed a fast tagger performing coarse-grained analyses[1] for English microtext with an accuracy around 92% [4]. Owoputi et al. [4] also train and test their tagger on the annotated data of Ritter et al. [5] and report an accuracy of around 90% on the 45 PTB tags plus the 4 (unambiguous) twitter-specific tags.

The English Twitter tagger of Owoputi et al. [4] mostly benefits from word clustering of unlabelled Twitter data. We follow this approach, using LDA word clusters and also test the impact of features from an automatically created dictionary as well as additional out-of-domain training data on the tagging accuracy.

3   Data

The data we use in our experiments was collected from Twitter[2] over a time period from July 2012 to February 2013. We used the Python Tweepy module[3]

---

[1] The Twitter POS Tagset of Gimpel et al. distinguishes 25 parts of speech. In comparison, the tagset used for the PTB includes 45 different tags and the original German Stuttgart-Tübingen Tagset (STTS) includes 54 different POS.
[2] https://de.twitter.com
[3] http://pythonhosted.org/tweepy/html

Table 1. Inter-annotator agreement for 2 annotators on the testset (507 tweets)

tweets	Fleiss κ
1-30	92.6
31-100	93.9
101-200	92.7
201-507	94.4

as an interface to the Twitter Search API[4] where we set the API language parameter to German. We collected a corpus of 12,782,097 tweets with a unique id. From those, we randomly selected 1,426 tweets (20,877 tokens). We manually annotated these tweets and divided them into a training set of 420 tweets (6,220 tokens), a development set of 500 tweets (7,232 tokens) and a test set of 506 tweets (7,425 tokens). Table 1 shows our inter-annotator agreement on the testset.

Comparing our German development set with English tweets using an equally-sized portion of the OCT27 data set [4], we observe a larger vocabulary size in the German data (German: 8,049, English: 7,201) as well as a higher number of hapax legomena (German: 7,061, English: 4,691). The tagging ambiguity in both data sets is very similar at 1,050 for German and 1,069 for English.

3.1 POS Tagset

The tagset we use for the annotation is an adapted version of the Stuttgart-Tübingen Tag Set (STTS) [6], a quasi-standard for German POS tagging which distinguishes 54 different tags. We follow Gimpel et al. [3] in providing additional tags for twitter-specific phenomena (hashtags, at-mentions, urls, emoticons). But note that we annotate hashtags and at-mentions which are used within in a phrase or sentence according to the STTS tag set (Example (1)). This is in contrast to Gimpel et al. [3] who do the same with hashtags but always annotate at-mentions as such, even when used as proper names within the sentence. This also contrasts with Ritter et al. [5] who do not have a tag for emoticons and who annotate all words marked with a # or an @ as well as all urls unambiguously with the tags HASHTAG, AT-MENTION, or URL, respectively, regardless of their function and position in the tweet. This detail is not trivial as urls and tokens marked with either a # or an @ do account for a substantial part of the corpus (in the case of our testset they amount to nearly 10% of all tokens).

(1)   **@swr3**/ADDRESS ich leider   nicht , aber **@mondmiri**/NE :)
       @swr3   I   regrettably not   , but **@mondmiri**   :)
       "@swr3 Not me, unfortunately, but @mondmiri :)

We also introduce a new tag for non-verbal comments (COMMENT). These are often expressed as non-inflected verb forms which are a frequent stilistic means

---

[4] https://dev.twitter.com/docs/api/1/get/search

Table 2. Extensions to the STTS (1-4: twitter-specific, 5-15: twitter and spoken language)

POS	description	example	translation
References			
1 URL	*URL*	http://t.co/LV3bTzAK	
2 HASHTAG	*hashtag*	#knp #News	
3 ADDRESS	*at-mention*	@konfus @me5594	
Non-verbal information			
4 EMO	*emoticon*	:) o_o :-P *_*	
5 COMMENT	*non-inflected forms*	*seufz*	*sigh*
	and other comments	*kopfschüttel*	*shake head*
Verbalised information			
6 PPER_ES	*amalgamated forms*	wie **ichs** mach	how **I_it** do
VVFIN_ES		etz **gibts** a bier	now **gives_it** a beer
KOUS_ES		**obs** morgen	**whether_it** tomorrow
...		regnet ...	rains ...
7 PTKFILL	*particle, filler*	Ich **äh** ich komme auch .	I **er** I come too .
8 PTK	*particle, unspecific*	**Ja** kommst Du denn auch ?	**PTC** come you then too ?
9 PTKREZ	*backchannel signal*	A: Ich komme auch . B: **Hm-hm** .	A: I come too . B: **Uh-huh** .
10 PTKONO	*onomatopoeia*	**Bum bum** .	**Boom boom** .
11 PTKQU	*question particle*	Du kommst auch . **Ne** ?	You come too . **No** ?
12 PTKPH	*placeholder particle*	Er hat **dings** hier .	He has **thingy** here .
13 XYB	*unfinished word, interruption or self-correction*	Ich **ko** # Ich komme **Diens** äh Mittwoch .	I **co** # I'll come **Tues** eh Wednesday .
14 XYU	*uninterpretable*	ehzxa	
punctuation			
15 $#	*unfinished utterance*	Ich ko #	I co #

in German comics [7] and computer-mediated communication. Other tokens annotated as COMMENT are complex phrases with a non-inflected head as in (2), complex phrases with a nominal head as in (3) or types of stage directions as in (4) which are used to "set the stage" and provide additional information which can not be communicated with the means of a conventional written-text message.

(2)   *ein-bisschen-Aufmerksamkeit-schenk*
      a-bit-attention-give

(3)   *neinkeinpobelsondernpiercing*
      no-no-bogy-but-piercing

(4)   *Trommelwirbel*
      drum-roll

We also follow Gimpel et al. [3] in using complex tags for amalgamated word forms (e.g. pronoun+*it* (5), verb+*it*, auxiliary+*it*, modal verb+*it* (5), subordinating conjunction+*it*) rather than letting the tokeniser deal with the problem.[5]

(5)   Wenn **sies** $_{PPER\_ES}$ nicht will   **kanns** $_{VMFIN\_ES}$ steffi haben
        If   she_it   not   wants can_it   steffi have
        "If she doesn't want it, Steffi can have it"

Other extensions are taken from a tagset developed for German spoken language [8] analysing different types of discourse particles and disfluencies (filled pauses, question tags, backchannel signals, unfinished words). Overall, we used a set of 65 different tags for annotating the 1,426 tweets.[6] Additional tags not included in the STTS are shown in Table 2. 10.5% of the tokens in the manually annotated tweets have been assigned a tag from the extended tagset.

Our proposal is different from the one of Beißwenger et al. [9] who, with reference to their functional and semantic similarity, group verbal and non-verbal elements as well as references into one new group of *interaction signs*, which is further subdivided into interjections, responsives, emoticons, interaction words, interaction templates and addressing terms. In contrast, our classification distinguishes between references which are not part of the actual sentence or phrase (URL, HASHTAG, ADDRESS), information on a non-verbal level, (usually) not integrated in the sentence or phrase (EMO, COMMENT) and verbalised information as part of the actual sentence(s) or phrase(s) in the tweet.

We consider the proposed tagset as preliminary work and would like to put it up for discussion in the NLP community, in order to advance the agreement on a widely accepted tagset for user-generated content from the web.

4   Experimental Setup

In our tagging experiments we use the CRFsuite package[7] [10] which provides a fast implementation of Conditional Random Fields [11] for sequence labelling. Below, we describe our feature set and the techniques used for adapting the tagger to the Twitter data. Unless stated otherwise, all results are reported on the development set.

---

[5] See Owoputi et al [4] for a discussion on this issue. Owoputi et al. argue that normalising the data and applying the same tokenisation as on the PTB would result in a "lossy translation" which would not do justice to many of the non-canonical phenomena such as markers of dialect or sociolect, and that text from social media should be analysed on its own terms.

[6] In theory, our tagset would have more tags, but 8 of the tags defined in the STTS did not occur in our data (e.g. postpositions and circumpositions).

[7] We run all our experiments with the default parameter setting (1st-order Markov CRF with dyad features, training method: limited memory BFGS).

Table 3. Feature set used in our experiments

feature	description
wrd	word form
len	word length
cap	first char == upper case?
hash	first char == #?
addr	first char == @?
upper	number of upper case in wrd
digit	number of digits in wrd
sym	number of non-char in wrd
pre N	first N char of wrd
suf N	last N char of wrd

4.1 Features

Table 3 shows the features used in our experiments. The *cap* feature is set to 1 if the first character of the word form is capitalised, and set to 0 otherwise. The *hash* and *addr* features are either set to 0 or 1, depending on the first character of the word being a # or a @, while the *upper*, *digit* and *sym* features count the number of upper case letters, digits or symbols in the input word form.

We experimented with different prefix and suffix sizes and settled on a size of up to 10 characters. We also converted all word forms to lowercase (but kept the features *cap* and *upper*). In addition, we use feature templates which generate new features of word ngrams where the input word form is combined with word forms from the context. We refer to this setting as exp01.

4.2 Word Clustering

Word clustering has been used for unsupervised and semi-supervised POS tagging, with considerable success [12–15]. It provides a way to obtain information on input tokens for which no annotated data is available.

For tagging English Twitter data, Owoputi et al. [15] apply Brown clustering, a hierarchical word clustering algorithm, to the unlabelled tweets. During clustering, each word is assigned a binary tree path. Prefixes of these tree paths are then used as new features for the tagger.

Chrupala [14, 16] proposes an alternative to Brown clustering, using Latent Dirichlet Allocation (LDA). LDA has two important advantages over Brown clustering. First, the LDA clustering approach is much more efficient in terms of training time. Second, LDA clustering produces soft, probabilistic word classes instead of the hard classes generated by the Brown algorithm, thus allowing one word to belong to more than one cluster. Chrupala treats words as documents and the contexts they occur in as the terms in the documents to induce a posterior distribution over K types and shows that his approach outperforms Brown clustering on many NLP tasks.

Table 4. Entries for *einen* in the automatically created HGC dictionary (ART: determiner; PIS: indefinite pronoun, substitutive; ADJA: adjective, attributive; NN: noun; VVFIN verb, finite; VVINF: verb, infinite)

frequency	word form	POS
410873	einen	ART
16550	einen	PIS
8679	einen	ADJA
438	einen	NN
160	einen	VVFIN
144	einen	VVINF

In exp02, we apply the LDA clustering approach using the software of Chrupala [14][8] to our unlabelled Twitter corpus. Before clustering, we normalise the data. Instead of word forms, we use lemma forms automatically generated by the tree-tagger (for unknown lemmas, we fall back to the word form). We convert all urls to <url>, hashtags, emoticons and at-mentions to <hashtag>, <emoticon> and <user>. Identical tweets are removed from the corpus. Our corpus contains 204,036,829 tokens and is much smaller than the one used in Owoputi et al. [15] which includes 847,000,000 tokens of Twitter microtext.

For each word form, we extract the five most probable classes predicted by the LDA clustering algorithm and add them to the feature set. To avoid a negative impact of peaked distributions, we use a threshold of 0.001 which a word class probability has to exceed for the word class to be included as a feature.

We test different settings for LDA, varying the threshold for the minimum number of occurrences for each word to be clustered, and induce clusters with 50 and 100 classes.

4.3 Learning from Out-of-Domain Data

In exp03 we explore whether training data from another domain can improve the accuracy on Twitter microtext. We use the German TiGer treebank [17], a corpus of newspaper text (888,578 tokens) annotated with the STTS tagset. We add different portions of the data to the training set and re-train the tagger. Please note that 17 of our tags are not part of the original STTS and thus do not occur in the out-of-domain data.

4.4 Adding Features from an Automatically Extracted Dictionary

In exp04 we take a self-training approach where we stack the tagger with POS tags obtained from an automatically created dictionary. The dictionary was harvested from the Huge German Corpus (HGC) [18], a collection of newspaper corpora from the 1990s with 253,706,256 tokens. We automatically POS tagged the HGC using the TreeTagger [19]. For each word form, we add the first 5 POS

---

[8] https://bitbucket.org/gchrupala/lda-wordclass/

Table 5. Results for different prefix/suffix sizes (01a) and for the same features with lowercase input (01b)

EXP	features	acc	EXP	acc.
01a	pre/suf 4	80.57	01b	81.71
	pre/suf 5	80.83		82.03
	pre/suf 6	80.87		82.18
	pre/suf 7	80.89		82.25
	pre/suf 8	80.97		82.31
	pre/suf 9	81.05		82.47
	pre/suf 10	81.15		**82.49**
	pre/suf 11	81.16		82.44
	pre/suf 12	81.13		82.42

tags assigned by the tagger as new features, ranked for frequency. To reduce noise, we only included POS which had been predicted at least 10 times for this particular word form.

As an example, consider the word form *einen*. For *einen*, our automatically created dictionary lists the entries in Table 4. We use the first 5 tags, ART, PIS, ADJA, NN and VVFIN, as additional features for the tagger.

5   Results

This section evaluates the different settings and domain adaptation techniques on the Twitter development data and discusses their impact on the number of OOV words in the data.

5.1   Impact of Prefix/Suffix Sizes

Table 5 shows results for different prefix/suffix sizes (exp01a). While a size of 4 seems to small to yield good results, the differences between the other settings are not significant. After converting all words to lowercase (but keeping the features *cap* and *upper*) we get a further improvement of around 1% (exp01b). The difference between exp01a (pre/suf 10) and exp01b (pre/suf 10, lowercase) is statistically significant (two-sided McNemar test, $p = 0.001$).

We obtained good results with a size of 10 and lowercase input (exp01b) and thus keep this setting for all following experiments.

5.2   LDA Word Cluster

Table 6 shows results for the different settings of our cluster approach. A cluster size of 50 gives slightly higher results than a size of 100, and a threshold of 10 gives us best results on the development set. Overall, we observe a significant improvement in accuracy of around 4 percentage points, confirming the adequacy of the clustering approach to overcome data sparseness.

Table 6. Impact of LDA word clustering on tagging accuracy on the development set

EXP	features	#train	cluster size	acc
02	lda-8	6220	50	85.97
	lda-10	6220	50	**86.29**
	lda-12	6220	50	86.15
	lda-20	6220	50	85.99
	lda-8	6220	100	85.99
	lda-10	6220	100	85.93
	lda-12	6220	100	85.84
	lda-20	6220	100	85.95

5.3 Out-of-Domain Data

Table 7 (exp03) shows results for adding portions of 100,000, 300,000, 500,000 and 800,000 tokens of annotated newspaper text from the TiGer treebank to the Twitter training set. The out-of-domain data gives us a further improvement of nearly 3 percentage points over exp02. Adding 500,000 tokens to the training data gives the best results. After adding even more data results slightly decrease. However, the difference between adding 300,000 and 500,000 tokens is not statistically significant.

5.4 HGC Dictionary

The success of the clustering approach again shows that our main problem is caused by the high number of unknown words. Following up on this, we test whether the features from the automatically created dictionary (exp04) can further mitigate this problem. Table 7 (exp03+04) shows that this method yields another small, but significant improvement over the best result from exp03 (two-sided McNemar test, $p < 0.001$).

6 Evaluation

Table 8 presents our final results for the different settings on the test set. Our basic feature set with lowercase input (exp01b) achieves an accuracy of 81.58.

Table 7. Adding different portions of newspaper text (03) and automatically created dictionaries from the HGC (04) to the training data

EXP	features	#train	acc
03	lda-10-50	106,220	88.58
	lda-10-50	306,220	88.99
	lda-10-50	506,220	89.11
	lda-10-50	806,220	88.54
04	lda-10-50 hgc	6,220	87.77
03+04	lda-10-50 hgc	306,220	**89.45**

Table 8. Final results on the test set

EXP	features	#train	dev	test
exp01	pre/suf 10 lc	6,220	82.49	81.58
exp02	lda-10-50	6,220	86.29	85.06
exp03	lda-10-50	306,220	88.99	88.00
exp03+04	lda-10-50 hgc	306,220	89.45	88.84

Adding the LDA word clusters as features results in a substantial boost of around 3.5% (exp02). The out-of-domain training data improves results by another 3 percentage points (exp03). Adding the features from the automatically created dictionary (exp03+04) gives us best results with 88.84 on the test set, which is significantly higher than exp03 (two-sided McNemar test, $p = 0.001$), but also substantially lower than the ones on the development set.

The lower results on the testset can partly be explained by a higher number of tokenisation errors, as we used the development set to improve the tokeniser. In the development set, 20 errors are due to wrong tokenisation while on the testset our tokeniser produces 61 errors.

To see what we have learned so far, we now have a look at the different settings in our experiments and evaluate their impact on the number of OOV words and on tagging accuracy.

6.1 Out-of-Vocabulary Words

Table 9 shows the impact of the different techniques on the number of OOV words in the development set. 2,684 word types from the development set are not included in the training set. After converting the word forms to lowercase, we still have 2,500 unknown words in the development set (exp01). Using features from LDA word clustering dramatically reduces the number of OOV words to 410 (exp03). Adding the out-of-domain training data further reduces the number of OOV words to 323. After adding the dictionary from the HGC, we have reduced the number of unknown words to 248.

To conclude, our efforts to address the OOV problem in the data resulted in a reduction of unknown words by a factor of ten. The remaining OOV words in the data are mostly compounds like "Weihnachtsschnäppchenangebote" *christmas bargain*, "Stadtfestprogrammhefte" *town fair programme brochure*,

Table 9. Impact of different settings on the number of OOV words (development set)

setting	OOV (types)	acc.
	2684	n.a.
lowercase	2500	82.21
+lda 10-50	410	86.29
+out-of-domain data	323	88.99
+ HGC	248	89.45

Table 10. Impact of adding a TiGer dictionary and annotated TiGer data on the tagging accuracy

EXP	features	#train	dev
exp04	lda 10-50 hgc	6,220	87.77
exp03+04	lda-10-50 hgc	306,220	**89.42**
exp05	lda hgc+tig.	6,220	87.82

"Donnerstagshandel" *thursday trade* or "Berufspudel" *professional poodle* and names of locations (Rotenburg-Wümme, Pfaffenhofen-Ilm, Simon-von-Utrecht-Straße). Also, creative inventions of new compounds as in "wer-hat-welchen-anime-mit-wem-wieweit-geschaut-dependency-graphen" *who-has-watched-which-anime-with-whom-for-how-long-dependency-graph* can be found.

We would like to know whether the substantial improvement we get when training on additional out-of-domain data is merely due to a reduction of OOV words or whether the data is able to provide the tagger with structural knowledge. To test this we extract a dictionary from the TiGer treebank, automatically annotated by the TreeTagger [19]. As the tagger was developed on the TiGer treebank, we do expect the automatic annotations to be of high quality. We do not use the gold annotations in TiGer because we want the features to be as similar to the ones from the HGC as possible.

Table 10 (exp05) shows that combining the dictionaries extracted from the HGC and TiGer does not result in a significant improvement over exp04 and yields substantially lower results than combining LDA word clusters, the HGC dictionary and the out-of-domain TiGer training data (exp03+04). This suggests that the tagger is able to learn important information from the out-of-domain data which cannot be provided by a simple dictionary.

6.2 Error Analysis

Our best tagging model produced 908 errors on the development set, 68 of which were caused by a confusion of different verb types, e.g. annotating an imperative

Table 11. The 10 most frequent errors made by the tagger (development set)

freq.	gold	predicted
15	PDS	ART
16	VVFIN	NN
20	ADV	NN
21	NE	HASH
22	VVFIN	VVINF
23	ADJD	NN
25	NN	HASH
29	ADJA	NN
75	NN	NE
133	NE	NN

verb form as a finite one and vice versa, or by predicting a finite verb (VVFIN) instead of an infinite one (VVINF). This error is also frequent on newspaper text, caused by a syncretism of the plural form of German finite verbs and the infinite verb form.

Table 11 lists the 10 most frequent errors made by the tagger on the development set. Among these is the false annotation of adjectives (ADJA, ADJD) as nouns (NN) or the confusion of definite determiners (ART) and demonstrative pronouns (PDS). The most frequent error made by the tagger was a mix-up of common nouns (NN) and proper names (NE).

Figure 1 shows the impact of the different settings on the number of errors made by the tagger. Most interestingly, while the features from the HGC do improve the overall tagging accuracy, they also result in a higher number of errors on nouns (NN), adverbial or predicative adjectives (ADJD), unfinished words (XYB), past participles (VVPP), verb particles (PTKVZ), indefinite pronouns (PIS, PIAT), interjections (ITJ) and finite modals (VMFIN).

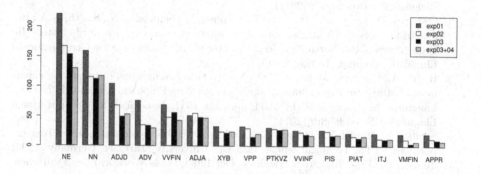

Fig. 1. Error reduction for individual POS tags for the most frequent error types

7 Conclusions

We presented the first work on fine-grained POS tagging for German Twitter data which, due to the richer morphology and semi-free word order in combination with case syncretism, constitutes a challenging test case for tagger adaptation. We extended the STTS for the annotation of user-generated content from the web and manually annotated 1,426 tweets (20,877 tokens). We showed that despite the significantly higher ratio of unknown words in the data, we are able to obtain tagging accuracies in the same range as the ones for English [5, 3, 4]. Crucially, our analysis is far more fine-grained than the one of Gimpel et al. [3] and will thus be of great value for the linguistic analysis of data from the social media.

Tokenisation of CMC is a challenging task which needs to be addressed as the tokenisation errors have a crucial impact on POS tagging accuracy. Also, using a larger amount of unlabelled Twitter data for learning word clusters might improve results. We will explore these issues in future work.

References

1. Foster, J.: "cba to check the spelling" investigating parser performance on discussion forum posts. In: Human Language Technologies: The 2010 Annual Conference of the North American Chapter of the Association for Computational Linguistics, HLT 2010, pp. 381–384. Association for Computational Linguistics, Stroudsburg (2010)
2. Foster, J., Wagner, J., Roux, J.L., Hogan, S., Nivre, J., Hogan, D., Genabith, J.V.: #hardtoparse: POS tagging and parsing the twitterverse. In: Proceedings of AAAI 2011 Workshop on Analysing Microtext (2011)
3. Gimpel, K., Schneider, N., O'Connor, B., Das, D., Mills, D., Eisenstein, J., Heilman, M., Yogatama, D., Flanigan, J., Smith, N.A.: Part-of-speech tagging for twitter: annotation, features, and experiments. In: Proceedings of the 49th Annual Meeting of the Association for Computational Linguistics: Human Language Technologies: Short Papers, HLT 2011, vol. 2, pp. 42–47. Association for Computational Linguistics, Stroudsburg (2011)
4. Owoputi, O., O'Connor, B., Dyer, C., Gimpel, K., Schneider, N., Smith, N.A.: Improved part-of-speech tagging for online conversational text with word clusters. In: Proceedings of the North American Chapter of the Association for Computational Linguistics Annual Meeting (2013)
5. Ritter, A., Clark, S., Mausam, E.O.: Named entity recognition in tweets: an experimental study. In: Proceedings of the Conference on Empirical Methods in Natural Language Processing, EMNLP 2011, pp. 1524–1534. Association for Computational Linguistics, Stroudsburg (2011)
6. Schiller, A., Teufel, S., Thielen, C.: Guidelines für das Tagging deutscher Textcorpora mit STTS. Technical report, IMS-CL. University Stuttgart, Germany (1995)
7. Teuber, O.: Fasel beschreib erwähn – Der Inflektiv als Wortform des Deutschen. Germanistische Linguistik 26(6), 141–142 (1998)
8. Rehbein, I., Schalowski, S.: Extending the STTS for the annotation of spoken language. In: Proceedings of KONVENS 2012, pp. 238–242 (2012)
9. Beißwenger, M., Ermakova, M., Geyken, A., Lemnitzer, L., Storrer, A.: A TEI schema for the representation of computer-mediated communication. Journal of the Text Encoding Initiative (3), 1–31 (2012)
10. Okazaki, N.: CRFsuite: a fast implementation of conditional random fields, CRFs (2007)
11. Lafferty, J.D., McCallum, A., Pereira, F.C.N.: Conditional random fields: Probabilistic models for segmenting and labeling sequence data. In: Proceedings of the Eighteenth International Conference on Machine Learning, ICML 2001, pp. 282–289. Morgan Kaufmann Publishers Inc., San Francisco (2001)
12. Biemann, C.: Unsupervised part-of-speech tagging employing efficient graph clustering. In: Proceedings of the 21st International Conference on Computational Linguistics and 44th Annual Meeting of the Association for Computational Linguistics: Student Research Workshop, COLING ACL 2006, pp. 7–12. Association for Computational Linguistics, Stroudsburg (2006)
13. Søgaard, A.: Simple semi-supervised training of part-of-speech taggers. In: Proceedings of the ACL 2010 Conference Short Papers, ACLShort 2010, pp. 205–208. Association for Computational Linguistics, Stroudsburg (2010)

14. Chrupala, G.: Efficient induction of probabilistic word classes with LDA. In: Proceedings of 5th International Joint Conference on Natural Language Processing, pp. 363–372. Asian Federation of Natural Language Processing, Chiang Mai (November 2011)
15. Owoputi, O., O'Connor, B., Dyer, C., Gimpel, K., Schneider, N.: Part-of-speech tagging for twitter: Word clusters and other advances. Technical Report CMU-ML-12-107. Carnegie Mellon University (2012)
16. Chrupala, G.: Hierarchical clustering of word class distributions. In: Proceedings of the NAACL-HLT Workshop on the Induction of Linguistic Structure, Montréal, Canada. Association for Computational Linguistics, pp. 100–104 (June 2012)
17. Brants, S., Dipper, S., Hansen, S., Lezius, W., Smith, G.: The TIGER treebank. In: Proceedings of the First Workshop on Treebanks and Linguistic Theories, pp. 24–42 (2002)
18. Fitschen, A.: Ein computerlinguistisches Lexikon als komplexes System. PhD thesis, Institut für Maschinelle Sprachverarbeitung der Universität Stuttgart (2004)
19. Schmid, H.: Improvements in part-of-speech tagging with an application to German. In: Proceedings of the ACL SIGDAT-Workshop (1995)

Data-Driven vs. Dictionary-Based Word n-Gram Feature Induction for Sentiment Analysis

Robert Remus[1] and Sven Rill[2,3]

[1] Natural Language Processing Group,
Department of Computer Science,
University of Leipzig, Germany
rremus@informatik.uni-leipzig.de
[2] Goethe University Frankfurt, Germany
[3] Institute of Information Systems,
University of Applied Sciences Hof,
Hof, Germany
srill@iisys.de

Abstract. We address the question which word n-gram feature induction approach yields the most accurate discriminative model for machine learning-based sentiment analysis within a specific domain: a purely data-driven word n-gram feature induction or a word n-gram feature induction based on a domain-specific or domain-non-specific polarity dictionary. We evaluate both approaches in document-level polarity classification experiments in 2 languages, English and German, for 4 analog domains each: user-written product reviews on books, DVDs, electronics and music. We conclude that while dictionary-based feature induction leads to large dimensionality reductions, purely data-driven feature induction yields more accurate discriminative models.

Keywords: Sentiment analysis, feature induction.

1 Introduction

Learning discriminative word n-gram models is a popular technique for solving several sentiment analysis subtasks [1], e.g. document-level polarity classification [2] or sentence-level subjectivity classification [3], as well as text classification in general [4, 5]. However, there are at least two competing approaches to word n-gram feature induction: *data-driven* feature induction, in which word n-gram features are extracted directly from the textual data, and *dictionary-based* feature induction, in which only those word n-grams are used as features, that appear in a pre-defined dictionary. While the former approach is independent of such often labor-intensive resources, these dictionaries allow the latter approach for a *feature selection* specifically tailored to a certain task, e.g. polarity classification [6, 7]. This feature selection may then may significantly reduce both noise and feature space size.

Either way, models that rely on word n-grams as features and that are trained on data that originates from a specific genre and a specific domain are generally

I. Gurevych, C. Biemann, and T. Zesch (Eds.): GSCL 2013, LNAI 8105, pp. 176–183, 2013.

highly genre- and domain-dependent [8–10]. This is because *genres*, e.g. newspaper articles or weblog posts, i.e. text categories based on external, non-linguistic criteria such as the intended audience, purpose and activity type [11], differ in their textual characteristics [12, 13]. In turn, *domains*, i.e. the subject area a certain newspaper article or weblog post deals with, differ in their vocabulary and in the way this vocabulary is used [14].

For the same reasons, a polarity dictionary, i.e. a lexical resource that lists word n-grams and their prior polarity [15], may be *domain-specific* or of more general nature, i.e. *domain-non-specific*. A domain-non-specific polarity dictionary contains word n-grams that bear an unambiguous prior polarity in many domains, e.g. "best" or "worst". In contrast, a domain-specific polarity dictionary additionally contains word n-grams that bear a clear prior polarity only within specific domains, e.g. "scary": whereas in most domains "scary" has a negative prior polarity, within the domains of horror books or horror movies it may have a positive prior polarity. Generally, words with domain-specific polarity, e.g. "small" or "large", "old" or "new" may be polar in one domain, but non-polar or of opposite polarity in another [16, 17].

In this work, we address the following question: Which word n-gram feature induction approach yields the most accurate discriminative model for machine learning-based sentiment analysis within a specific domain: (i) a purely data-driven word n-gram feature induction, (ii) a word n-gram feature induction based on a domain-specific polarity dictionary or (iii) a word n-gram feature induction based on a domain-non-specific polarity dictionary? To the best of our knowledge, such a study has not been carried out before.

This paper is structured as follows: In the next section we shortly describe the data-driven and dictionary-based word n-gram feature induction approaches we compare. In Section 3 we evaluate them in a common sentiment analysis subtask, document-level polarity classification, in 2 languages, English and German, and 4 domains each. Finally, we draw conclusions and point out possible directions for future work in Section 4.

2   Word n-Gram Feature Induction

A word n-gram *feature induction*, sometimes also referred to as feature extraction, induces features on textual data, e.g. a document or a sentence, based on a set of word n-grams. Thereby, the textual data is represented in a feature space, usually encoding the presence or absence of these word n-grams, or other measures thereof, e.g. their frequency [2]. The word n-grams to be used as features may either be chosen using a data-driven approach or a dictionary.

In a purely *data-driven* feature induction, for every word n-gram *type* present in the textual data a feature is created, i.e. the feature space size equals the word n-gram vocabulary size. Typically, word uni-, bi- and/or trigrams are used as features for text classification. Note that such a data-driven feature induction utilizes no prior knowledge of the meaning or importance of certain word n-grams. Thus a purely data-driven feature induction is implicitly domain-specific, as it lets the data speak for themselves when learning a model.

In a *dictionary-based* feature induction, only those word n-gram types extracted from the data are used as features, that also appear in a pre-defined dictionary. Therefore, dictionary-based feature induction may be seen as combined feature induction and feature selection, as only a subset of all word n-gram types present in the textual data are used as features. A dictionary-based feature induction is domain-non-specific if it utilizes a domain-non-specific dictionary, or domain-specific when it utilizes a domain-specific dictionary. The construction of such dictionaries may vary from task to task.

3   Evaluation   •

We evaluate the word n-gram feature induction approaches described in Section 2 in a common sentiment analysis subtask: document-level polarity classification. Our setup for all experiments is as follows: For sentence segmentation and tokenization we use *OpenNLP*[1]. As classifiers we employ Support Vector Machines (SVMs) [18] as implemented by *LibSVM*[2] using a linear kernel with their cost factor C set to 2.0 without any further optimization. SVMs were chosen because (i) it has been shown previously that they exhibit superior classification power in polarity classification experiments using word n-grams [2] and therefore (ii) nowadays SVMs are a common choice for sentiment analysis classification subtasks and other text classifications in general [19]. As features we use word n-grams chosen either by a dictionary-based feature induction using the polarity dictionaries described below, or combinations of word uni-, bi- and trigrams chosen by a purely data-driven feature induction. We simply encode the word n-grams' presence or absence. Dictionary-based feature induction aside, we perform no further feature selection; neither stop words nor punctuation characters are removed.

All binary classification experiments are construed as 10-fold cross validations. In each fold 9/10th of the available data are used for training, the remaining 1/10th is used for testing. Training and testing data never overlap. As performance measure we report accuracy A. The level of statistical significance is determined by *stratified shuffling*, an approximate randomization test [20] run with $2^{20} = 1,048,576$ iterations as recommended by [21].

As gold standard for English-language experiments we use [22]'s *Multi-domain Sentiment Dataset v2.0*[3] (MDSD), that contains star-rated product reviews of various domains from English Amazon. We chose 4 domains: books, dvd, electronics and music. For these domains a pre-selected, balanced amount of 1,000 positive and 1,000 negative reviews is available. [22] consider reviews with more than 3 stars positive, less than 3 stars negative and omit 3-star reviews; so do we. As a gold standard for German-language experiments we randomly extract a balanced amount of 1,000 positive (4 or 5 stars) and 1,000 (1 or 2 stars) negative reviews from the 1.5 million German-language user-written product reviews

---

[1] http://opennlp.apache.org
[2] http://www.csie.ntu.edu.tw/~cjlin/libsvm/
[3] http://www.cs.jhu.edu/~mdredze/datasets/sentiment/

described in Section 3.1 for each of the 4 domains: Bücher (books), Film (DVDs, movies), Elektronik (electronics) and Musik (music).

As domain-non-specific polarity dictionaries we use the English-language *Subjectivity Lexicon* [23] (SL) and *SentiWordNet* v3.0.0 [24, 25] (SWN), as well as the German-language *GermanPolarityClues* [6] (GPC) and *SentiWS* (SWS) [26]. The domain-specific polarity dictionaries are acquired as described in the following section.

3.1 Constructing Domain-Specific Dictionaries

We construct domain-specific *Sentiment Phrase Lists* (SePLs), one per domain and language as described in [27, 28][4]. A SePL is based on user-written product reviews. The basic idea is that polarity or sentiment expressed by a review's star rating is strongly correlated with the sentiment expressed by polarity-bearing words in the review's title. We extract these polar words from the review's title and derive their polarity strengths, which are on a scale between -1 (very negative) and $+1$ (very positive). Contrary to [27, 28], we require a phrase to appear only at least 5 times, instead of 10 times. This is due to the smaller amount of available reviews per domain and language (cf. Table 1) as opposed to the original SePL which is constructed using reviews of various domains.

We extract single *polar words* (e.g. "good") and *polar phrases* (e.g. "very good", "not good"), i.e. word n-grams ($n \geq 1$). Polar phrases contain at least one polar word. In addition, they may contain *valence shifters* [29] and/or *negations* [30]. Including valence shifters and negations into the dictionary as parts of polar phrases has the advantage that their *sentiment composition*, which exhibits many exceptions, has not to be taken care of during application of the dictionary. E.g. the negation "not" does not always flip a word's polarity strength: "good" has a positive polarity strength, its negation "not good" has a negative polarity strength. But for other words, a negation only shifts the polarity strength from strong to weak or even neutral. E.g. "perfect" has a large positive polarity strength, whereas "not perfect" has a lower positive polarity strength, but is still positive. Thus, not including valence shifters and negations into a dictionary requires a sophisticated treatment of them during application of the dictionary, as discussed in [31–34].

To construct a domain-specific SePL, a review data set of this particular domain is necessary. For the English-language SePLs we use the aforementioned MDSD. For the German-language SePLs we compiled our own data set: we collected about 1.5 million reviews from German Amazon. To ensure that only German-language reviews are used, a language detection was performed using the *Language Detection Library for Java*[5]. Table 1 provides an overview of the English- and German-language data sets as well as the extracted polar words and phrases of the domain-specific SePLs.

---

[4] The domain-specific SePLs are available upon request from the authors.
[5] http://code.google.com/p/language-detection/

Table 1. Overview of the English- and German-language data sets as well as the number of extracted polar phrases of the domain-specific SePLs

| Domain | number of reviews | | | | | | number of phrases |
	1-star	2-star	3-star	4-star	5-star	total	
books	67,362	55,510	0	216,225	628,118	967,215	7,868
dvd	9,765	7,688	0	31,294	74,997	123,744	1,168
electronics	3,427	1,621	0	5,511	12,450	23,009	349
music	7,826	6,811	0	36,816	121,227	172,680	1,541
Bücher	31,629	25,650	39,323	72,729	223,576	392,907	2,642
Film	19,442	12,604	17,810	29,625	103,196	182,677	1,435
Elektronik	19,759	12,486	17,645	44,745	149,807	244,442	1,297
Musik	6,781	5,299	8,077	15,847	83,086	119,090	967

3.2 Results

Table 2 shows evaluation results using the different feature induction approaches on English-language MDSD, Table 3 shows evaluation results on our German-language data set.

Table 2. Accuracy of the data-driven and dictionary-based feature induction approaches and their average number of induced features for the English-language MDSD

| Domain | Dictionary-based | | | Purely data-driven | | |
	SePL	SL	SWN	{uni}	{uni, bi}	{uni, bi, tri}
books	71.05	71.35	74.55	77	79	78.65
dvd	72.7	75.4	76.45	78.35	79.65	79.25
electronics	73.6	76.7	76.4	77.6	82.1	81.65
music	69.45	74.85	72.6	74.1	77.05	77.65
average	71.7	74.58	75	76.76	79.45	79.3
#(features)	2,732	6,443	147,306	23,364	149,171	357,719

On English-language MDSD, data-driven feature induction ($A = 76.8$ to $A = 79.5$) clearly outperforms feature induction based on domain-specific SePLs ($A = 71.7$) and domain-non-specific SL ($A = 74.6$) and SWN ($A = 75$) across all domains. The difference between data-driven feature inductions and dictionary-based feature inductions is always statistically significant ($p < 0.05$). Surprisingly, the feature induction based on domain-specific SePLs performs worse than the one based on domain-non-specific SL and SWN. Partially, this is due to (i) SePLs smaller *coverage* and (ii) SVM's ability to handle superfluous features.

Similarly, on the German-language data set, the data-driven feature induction ($A = 80.7$ to $A = 81.6$) clearly outperforms feature induction based on domain-specific SePLs ($A = 72.6$) and domain-non-specific SWS ($A = 73.1$) and GPC

Table 3. Accuracy of the data-driven and dictionary-based feature induction approaches and their average number of induced features for the German-language data set

Domain	Dictionary-based			Purely data-driven		
	SePL	SWS	GPC	{uni}	{uni, bi}	{uni, bi, tri}
books	74.95	73.05	76.05	79.8	81.1	80.4
dvd	70.5	70.2	73.15	78.9	79.9	79.35
electronics	73.65	75.85	77.8	82.4	83.3	82.35
music	71.1	73.25	78.1	82	82.05	80.8
average	72.55	73.09	76.28	80.78	81.59	80.73
#(features)	1,585	3,462	10,141	32,506	172,879	382,204

($A = 76.3$) across all domains. Again, the difference between data-driven feature inductions and dictionary-based feature inductions is always statistically significant ($p < 0.005$). Again, the feature induction based on domain-specific SePLs generally performs worse than the one based on domain-non-specific SWS and GPC.

Despite their inferior accuracy, feature induction based on domain-specific SePLs is superior with regard to the resulting *feature space*: it is on average 9 to 241 times smaller than that of any data-driven feature induction; it is also 2 to 54 times smaller than that of any feature induction based on domain-non-specific dictionaries.

4   Conclusions and Future Work

We conclude that for the 2 languages and 4 domains we experimented with (i) a purely data-driven word n-gram feature induction yields more accurate models than any polarity dictionary-based word n-gram feature induction; (ii) word n-gram feature induction based on domain-non-specific polarity dictionaries yields more accurate models than feature induction based on domain-specific polarity dictionaries; (iii) dictionary-based word n-gram feature induction still provides a viable alternative to purely data-driven feature induction, particularly in environments with strong restrictions of memory or computing power.

Future work includes extending our study to other sentiment analysis subtasks, e.g. sentence-level polarity classification. Furthermore, we will increase the coverage of domain-specific SePLs, (i) by using larger review data sets for dictionary acquirement and (ii) by including verbs into SePL, which currently only contains adjectives and nouns.

References

1. Pang, B., Lee, L.: Opinion mining and sentiment analysis. Foundations and Trends in Information Retrieval 2(1-2), 1–135 (2008)
2. Pang, B., Lee, L., Vaithyanathan, S.: Thumbs up? sentiment classification using machine learning techniques. In: Proceedings of the 40th Annual Meeting of the Association for Computational Linguistics (ACL), pp. 79–86 (2002)

3. Wiebe, J., Wilson, T., Bruce, R., Bell, M., Martin, M.: Learning subjective language. Computational Linguistics 30(3), 277–308 (2004)
4. Lewis, D.: Feature selection and feature extraction for text categorization. In: Proceedings of the Workshop on Speech and Natural Language, pp. 212–217 (1992)
5. Sebastiani, F.: Machine learning in automated text categorization. ACM Computing Surveys (CSUR) 34(1), 1–47 (2002)
6. Waltinger, U.: GermanPolarityClues: A lexical resource for German sentiment analysis. In: Proceedings of the 7th International Conference on Language Resources and Evaluation (LREC), pp. 1638–1642 (2010)
7. Waltinger, U.: An empirical study on machine learning-based sentiment classification using polarity clues. Web Information Systems and Technologies 75(4), 202–214 (2011)
8. Sekine, S.: The domain dependence of parsing. In: Proceedings of the 5th Conference on Applied Natural Language Processing (ANLP), pp. 96–102 (1997)
9. Escudero, G., Màrquez, L., Rigau, G.: An empirical study of the domain dependence of supervised word sense disambiguation systems. In: Proceedings of Joint Conference on Empirical Methods in Natural Language Processing (EMNLP) and Very Large Corpora (VLC), pp. 172–180 (2000)
10. Wang, D., Liu, Y.: A cross-corpus study of unsupervised subjectivity identification based on calibrated EM. In: Proceedings of the 2nd Workshop on Computational Approaches to Subjectivity and Sentiment Analysis (WASSA), pp. 161–167 (2011)
11. Lee, D.: Genres, registers, text types, domains, and styles: Clarifying the concepts and navigating a path through the bnc jungle. Language Learning & Technology 5(3), 37–72 (2001)
12. Bank, M., Remus, R., Schierle, M.: Textual characteristics for language engineering. In: Proceedings of the 8th International Conference on Language Resources and Evaluation (LREC), pp. 515–519 (2012)
13. Remus, R., Bank, M.: Textual characteristics of different-sized corpora. In: Proceedings of the 5th Workshop on Building and Using Comparable Corpora (BUCC), pp. 156–160 (2012)
14. Remus, R.: Domain adaptation using domain similarity- and domain complexity-based instance selection for cross-domain sentiment analysis. In: Proceedings of the 2012 IEEE 12th International Conference on Data Mining Workshops (ICDMW 2012), Workshop on Sentiment Elicitation from Natural Text for Information Retrieval and Extraction (SENTIRE), pp. 717–723 (2012)
15. Wilson, T., Wiebe, J., Hoffmann, P.: Recognizing contextual polarity: An exploration of features for phrase-level sentiment analysis. Computational Linguistics 35(3), 399–433 (2009)
16. Fahrni, A., Klenner, M.: Old wine or warm beer: Target-specific sentiment analysis of adjectives. In: Proceedings of the Symposium on Affective Language in Human and Machine, AISB Convention, pp. 60–63 (2008)
17. Wu, Y., Jin, P.: SemEval-2010 task 18: Disambiguating sentiment ambiguous adjectives. In: Proceedings of the 5th International Workshop on Semantic Evaluation (SemEval), pp. 81–85 (2010)
18. Vapnik, V.: The Nature of Statistical Learning. Springer, New York (1995)
19. Joachims, T.: Text categorization with support vector machines: Learning with many relevant features. In: Nédellec, C., Rouveirol, C. (eds.) ECML 1998. LNCS, vol. 1398, pp. 137–142. Springer, Heidelberg (1998)
20. Noreen, E.: Computer Intensive Methods for Testing Hypothesis – An Introduction. John Wiley and Sons, Inc. (1989)

21. Yeh, A.: More accurate tests for the statistical significance of result differences. In: Proceedings of the 18th International Conference on Computational Linguistics (COLING), pp. 947–953 (2000)
22. Blitzer, J., Dredze, M., Pereira, F.: Biographies, bollywood, boom-boxes and blenders: Domain adaptation for sentiment classification. In: Proceedings of the 45th Annual Meeting of the Association for Computational Linguistics (ACL), pp. 440–447 (2007)
23. Wilson, T., Wiebe, J., Hoffmann, P.: Recognizing contextual polarity in phrase-level sentiment analysis. In: Proceedings of the Conference on Human Language Technology (HLT) and the Conference on Empirical Methods in Natural Language Processing (EMNLP), pp. 347–354 (2005)
24. Esuli, A., Sebastiani, F.: SentiWordNet: A publicly available lexical resource for opinion mining. In: Proceedings of the 5th International Conference on Language Resources and Evaluation (LREC), pp. 417–422 (2006)
25. Baccianella, S., Esuli, A., Sebastiani, F.: SentiWordNet 3.0: An enhanced lexical resource for sentiment analysis and opinion mining. In: Proceedings of the 7th International Conference on Language Resources and Evaluation (LREC), pp. 2200–2204 (2010)
26. Remus, R., Quasthoff, U., Heyer, G.: SentiWS – a publicly available German-language resource for sentiment analysis. In: Proceedings of the 7th International Conference on Language Resources and Evaluation (LREC), pp. 1168–1171 (2010)
27. Rill, S., Scheidt, J., Drescher, J., Schütz, O., Reinel, D., Wogenstein, F.: A generic approach to generate opinion lists of phrases for opinion mining applications. In: Proceedings of the 1st International Workshop on Issues of Sentiment Discovery and Opinion Mining, WISDOM (2012)
28. Rill, S., Adolph, S., Drescher, J., Reinel, D., Scheidt, J., Schütz, O., Wogenstein, F., Zicari, R., Korfiatis, N.: A phrase-based opinion list for the german language. In: Proceedings of the 1st Workshop on Practice and Theory of Opinion Mining and Sentiment Analysis (PATHOS), pp. 305–313 (2012)
29. Polanyi, L., Zaenen, A.: Contextual Valence Shifters. In: Computing Attitude and Affect in Text: Theory and Applications. The Information Retrieval Series, vol. 20, pp. 1–9. Springer, Dordrecht (2006)
30. Wiegand, M., Balahur, A., Roth, B., Klakow, D., Montoyo, A.: A survey on the role of negation in sentiment analysis. In: Proceedings of the 2010 Workshop on Negation and Speculation in Natural Language Processing (NeSp-NLP), pp. 60–68 (2010)
31. Choi, Y., Cardie, C.: Learning with compositional semantics as structural inference for subsentential sentiment analysis. In: Proceedings of the 13th Conference on Empirical Methods in Natural Language Processing (EMNLP), pp. 793–801 (2008)
32. Klenner, M., Petrakis, S., Fahrni, A.: Robust compositional polarity classification. In: Proceedings of the 7th International Conference on Recent Advances in Natural Language Processing (RANLP), pp. 180–184 (2009)
33. Liu, J., Seneff, S.: Review sentiment scoring via a parse-and-paraphrase paradigm. In: Proceedings of the 14th Conference on Empirical Methods in Natural Language Processing (EMNLP), pp. 161–169 (2008)
34. Moilanen, K., Pulman, S.: Sentiment composition. In: Proceedings of the 6th International Conference on Recent Advances in Natural Language Processing (RANLP), pp. 378–382 (2007)

Pattern-Based Distinction of Paradigmatic Relations for German Nouns, Verbs, Adjectives

Sabine Schulte im Walde and Maximilian Köper

Institut für Maschinelle Sprachverarbeitung, Universität Stuttgart, Germany

Abstract. This paper implements a simple vector space model relying on lexico-syntactic patterns to distinguish between the paradigmatic relations *synonymy, antonymy* and *hypernymy*. Our study is performed across word classes, and models the lexical relations between German nouns, verbs and adjectives. Applying *nearest-centroid classification* to the relation vectors, we achieve a precision of 59.80%, which significantly outperforms the majority baseline (χ^2, p<0.05). The best results rely on large-scale, noisy patterns, without significant improvements from various pattern generalisations and reliability filters. Analysing the classification shows that (i) antonym/synonym distinction is performed significantly better than synonym/hypernym distinction, and (ii) that paradigmatic relations between verbs are more difficult to predict than paradigmatic relations between nouns or adjectives.

1 Introduction

Paradigmatic relations (such as synonymy, antonymy and hypernymy, cf. [1]), are notoriously difficult to distinguish because the first-order co-occurrence distributions of the related words tend to be very similar across the relations. For example, with regard to the sentence *The boy/girl/person loves/hates the cat,* the nominal co-hyponyms *boy, girl* and their hypernym *person* as well as the verbal antonyms *love* and *hate* occur in identical contexts, respectively. Accordingly, while there is a rich tradition on identifying paradigmatically related word pairs in isolation (cf. [2–4] on synonymy, [5–7] on antonymy and [8–10] on hypernymy, among many others), there is little work that has addressed the distinction between two or more paradigmatic relations (such as [11–13] on distinguishing synonyms from antonyms).

The current study applies a simple vector space model to the distinction of paradigmatic relations in German, across the three word classes of nouns, verbs and adjectives. The vector space model is generated in the tradition of lexico-syntactic patterns: we rely on the linear sequences between two simplex words (representing synonyms, antonyms or hypernyms) as vector features in order to predict the lexical semantic relation between the two words. Our hope is that the vector space models using such patterns will unveil differences between the semantic relation pairs. For example, intuitively 'und' (*and*) should be a 1-word pattern to connect synonyms rather than antonyms, while 'oder' (*or*) should be a 1-word pattern to connect antonyms rather than synonyms. The

I. Gurevych, C. Biemann, and T. Zesch (Eds.): GSCL 2013, LNAI 8105, pp. 184–198, 2013.

pattern-based approach to distinguish lexical semantic relations has first been proposed by [8] to identify *noun hypernyms*; subsequent prominent pattern-based approaches are [14, 15] who identified *noun meronyms*; [16] on *noun causality*; [17] on *verb similarity, strength, antonymy, enablement, happens-before*; [18] on *noun hypernymy, meronymy, succession, reaction, production*; and [19] on *noun relational analogies*. (See Section 2 for more details on related work.) Our main questions with regard to the study can be summarised as follows.

- Can lexico-syntactic patterns distinguish between paradigmatic relations?
- Which relations are more difficult to distinguish than others?
- What are the differences across word classes?

2   Related Work

Although there are not many approaches in Computational Linguistics that explicitly addressed the distinction of paradigmatic semantic relations, there is a rich tradition on either synonyms or antonyms or hypernyms. Prominent work on identifying **synonyms** has been provided by Edmonds who employed a co-occurrence network and second-order co-occurrence (e.g., [20–22, 2]), and Curran who explored word-based and syntax-based co-occurrence for thesaurus construction (e.g., [23, 3]). [24] presented two methods (using patterns vs. bilingual dictionaries) to identify synonyms among distributionally similar words; [4] compared a standard distributional approach against cross-lingual alignment; [25] defined a vector space model for word meaning in context, to identify synonyms and the substitutability of verbs. Most computational work addressing **hypernyms** was performed for nouns, cf. the lexico-syntactic patterns by [8] and an extension of the patterns by dependency paths [10]. [26, 27] represent systems that identify hypernyms in distributional spaces. Examples of approaches that addressed the automatic construction of a hypernym hierarchy (for nouns) are [28, 9, 29–31]. Hypernymy between verbs has been addressed by [32–34]. Comparably few approaches have worked on the automatic induction of **antonyms**. A cluster of approaches in the early 90s tested the co-occurrence hypothesis, e.g., [35, 36, 5]. In recent years there have been approaches to antonymy that were driven by text understanding efforts, or being embedded in a larger framework to identify contradiction [37, 6, 7, 38].

Among the few approaches that distinguished *between* paradigmatic semantic relations we only know about systems addressing **synonyms vs. antonyms**. [24] implemented a similarity measure to retrieve distributionally similar words for constructing a thesaurus. They used a post-processing step to filter out any words that appeared with the patterns 'from X to Y' or 'either X or Y' significantly often, as these patterns usually indicate opposition rather than synonymy. [11] tackled the task within a pattern-based approach (see below). A recent study by [13], whose main focus was on the identification and ranking of opposites, also discussed the task of synonym/antonym distinction as a specific application of their findings.

Regarding pattern-based approaches to identify and distinguish lexical semantic relations in more general terms, [8] was the first to propose lexico-syntactic patterns as empirical pointers towards relation instances. Her goal was to identify pairs of nouns where one of the nouns represented the hypernym of the other. She started out with a handful of manual patterns such as

NP_i {, NP_j} * {,} and other NP_k

that were clear indicators of the lexical relationship (in this case with NP_i and NP_j representing hyponyms of NP_k), and used bootstrapping to alternately (i) find salient instances on the basis of the patterns, and (ii) rely on the enlarged set of pair instances to identify more salient patterns that are indicators of the relationship. Hearst demonstrated the success of her approach by comparing the retrieved noun pairs with WordNet lexical semantic relation pairs.

Girju [16] distinguished pairs of nouns that are in a causal relationship from those that are not. Differently to Hearst, she only relied on a single pattern

NP_i verb NP_k

that represented a salient indicator of causation between two nouns (with NP_i representing the cause and NP_k the effect) but at the same time was a very ambiguous pattern. Girju used a Decision Tree on 683 noun pairs and predicted the existence of a causal relation with a precision of 73.91% and a recall of 88.69%; in addition, she applied the causation prediction to question answering and reached a significant improvement. In [15], Girju and colleagues extended the lexical relation work to part–whole relations, applying a supervised, knowledge-intensive approach, mainly relying on WordNet and semantically annotated corpora. As in the earlier work, the task was to distinguish positive and negative relation instances. While they reached an f-score of 82.05%, they noted that many of the lexico-syntactic patterns were highly ambiguous (i.e., depending on the context they indicated different relationships).

[17] were the first to apply pattern-based relation extraction to verbs. For five non-disjoint lexical semantic relations (similarity, strength, antonymy, enablement, happens-before) they manually defined patterns and then queried Google to estimate joint pair–pattern frequencies for WordNet pairs as well as verb pairs generated by DIRT [39]. The accuracy for predicting whether a certain pair undergoes a certain semantic relationship varied between 50% and 100%, for relation set sizes of 2–41.

[18] developed Espresso, a weakly-supervised system that exploits patterns in large-scale web data. Similarly to [15], they used generic patterns, but relied on a bootstrapping cycle combined with reliability measures, rather than manual knowledge resources. Espresso worked in three phases: pattern induction, pattern selection and instance extraction. Starting with seed instances for the lexical semantic relations, the bootstrapping cycle iteratively induced patterns and new relation instances by web queries. Each induction step was combined with filtering out the least salient patterns/instances by reliability measures. The approach was applied to five noun-noun lexical semantic relations (hypernymy, meronymy, succession, reaction, production) and reached accuracy values between 49% and 91%, depending on the data and the relationship.

The work by Turney also includes approaches to extract and distinguish word pairs with regard to their lexical semantic relation. He developed a framework called *Latent Relational Analysis (LRA)* [40, 41, 19] that relied on corpus-based patterns between words in order to model relational similarity, i.e., similarity between word pairs *A:B::C:D* such that *A is related to B as C is related to D*. In his framework, a vector space model was populated with word pairs as the targets and patterns as the pair features. The patterns were derived from web corpora, and the cosine was used to measure the relational similarity between two word pairs. Turney applied a range of modifications to his basic setup, including a step-wise generalisation of the patterns by wild-cards instead of specific word types; extension of target pairs by synonyms to the words within a pair, as determined by Lin's thesaurus [42]; feature reduction by Singular Value Decomposition; etc. LRA has been applied to predict analogies in semantic relation pairs, to classify noun-modifier pairs according to the noun-noun semantic relation; to identify TOEFL synonyms; to answer SAT questions; to distinguish synonyms and antonyms; among others.

3   Paradigmatic Relation Datasets

The dataset of paradigmatic relations used in our research has been collected independently of the specific classification task in this paper. Based on a selection of semantic relation targets across the three word classes nouns, verbs and adjectives, we collected antonyms, synonyms and hypernyms for these targets via crowdsourcing. The following steps describe the creation of the dataset in more detail.

1. **Target Source, Semantic Classes and Senses:** We selected GermaNet[1] [43–45] as the source for our semantic relation targets. GermaNet is a lexical-semantic taxonomy for German that defines semantic relations between word senses, in the vein of the English WordNet [46]. Relying on GermaNet version 6.0 and the respective *JAVA API*, we generated lists of all nouns, verbs and adjectives, according to their semantic class (as represented by the file organisation), and also extracted the number of senses for each lexical item.

2. **Target Frequencies:** Relying on the German web corpus *sdeWaC* (version 3), we extracted corpus frequencies for all lexical items in the GermaNet files, if available. The *sdeWaC* corpus [47] is a cleaned version of the German web corpus *deWaC* created by the *WaCky* group [48]. It contains approx. 880 million words with lemma and part-of-speech annotations [49] and can be downloaded from http://wacky.sslmit.unibo.it/.

3. **Target Selection:** Using a stratified sampling technique, we randomly selected 99 nouns, 99 adjectives and 99 verbs from the GermaNet files. The random selection was balanced for

---
[1] www.sfs.uni-tuebingen.de/lsd/

(a) the **size of the semantic classes**,[2] accounting for the 16 semantic adjective classes and the 23 semantic classes for both nouns and verbs;

(b) **three polysemy classes** according to the number of GermaNet senses: I) monosemous, II) two senses and III) more than two senses;

(c) **three frequency classes** (type frequency in sdeWaC): I) *low* (200–2,999), II) *mid* (3,000–9,999) and III) *high* ($\geq$10,000).

The total number of 99 targets per word class resulted from distinguishing 3 sense classes and 3 frequency classes, $3 \times 3 = 9$ categories, and selecting 11 instances from each category, in proportion to the semantic class sizes.

4. **Semantic Relation Generation:** An experiment hosted by Amazon Mechanical Turk (AMT)[3] collected synonyms, antonyms and hypernyms for each of our 3×99 targets. For each word class and semantic relation, the targets were distributed randomly over 9 batches including 9 target each. In order to control for spammers, we in addition included two German fake words into each of the batches, in random positions of the batches. If participants did not recognise the fake words, all of their data were rejected. We asked for 10 participants per target and relation, resulting in 3 word classes $\times$ 99 targets $\times$ 3 relations $\times$ 10 participants $= 8,910$ target–response pairs. Table 1 shows some examples of the generated target–response pairs across the word classes and relations. The examples are accompanied by the *strength* of the responses, i.e., the number of participants who provided the response.

Table 1. Examples of target–response pairs across word classes and semantic relations

	ANT		SYN		HYP	
NOUN	*Bein/Arm* (leg/arm)	10	*Killer/Mörder* (killer)	8	*Ekel/Gefühl* (disgust/feeling)	7
	Zeit/Raum (time/space)	3	*Gerät/Apparat* (device)	3	*Arzt/Beruf* (doctor/profession)	5
VERB	*verbieten/erlauben* (forbid/allow)	10	*üben/trainieren* (practise)	6	*trampeln/gehen* (lumber/walk)	6
	setzen/stehen (sit/stand)	4	*setzen/platzieren* (place)	3	*wehen/bewegen* (wave/move)	3
ADJ	*dunkel/hell* (dark/light)	10	*mild/sanft* (smooth)	9	*grün/farbig* (green/colourful)	5
	heiter/trist (cheerful/sad)	2	*bekannt/vertraut* (familiar)	4	*heiter/hell* (bright/light)	1

We decided in favour of this very specific dataset and against directly using the GermaNet relations, for the following reason. Although GermaNet aims to include examples of all three relation types for each of the three parts-of-speech (nouns, verbs, adjectives), coverage of these can be low in places, as depending on the part-of-speech some semantic relations apply more naturally than others [50]. For example, the predominant semantic relation for nouns is hypernymy, whereas the predominant semantic relation for adjectives is antonymy. As a result, GermaNet does not always provide all three relations with regard to a specific lexical unit.

---

[2] For example, if an adjective GermaNet class contained a total of 996 word types, and the total number of all adjectives over all semantic classes was 8,582, and with 99 stimuli collected in total, we randomly selected $99 * 996/8,582 = 11$ adjectives from this semantic class.

[3] https://www.mturk.com

4   Experiments

The goal of our experiments was to distinguish between the three paradigmatic relations *antonymy, synonymy, hypernymy*. The following subsections describe the setup of the experiments (Section 4.1) and the results (Section 4.2).

4.1   Setup

Dataset: The experiments rely on a subset of the collected pairs as described in the previous section, containing those target–response pairs that were provided at least twice (to ensure reliability) and without ambiguity[4] between the relations. Table 2 shows the distribution of the target–response pairs across classes and relations. The target–relation pairs were randomly divided into 80% training pairs and 20% test pairs with regard to each class–relation combination.

Table 2. Target–response pairs

	ANT	SYN	HYP
NOUN	95	90	97
VERB	75	76	74
ADJ	62	62	61

In addition to using this dataset, the overall best experiments were performed on a variant that investigated the influence of polysemy among the targets and responses. We relied on the same dataset but distinguished between monosemous vs. polysemous target–response pairs. I.e., we divided the training pairs and the test pairs into two sets for each class–relation combination, one containing only pairs where both the target and the response were monosemous, and one containing only pairs where either the target or the response was polysemous, according to the definitions in GermaNet. (The third case, that both target and response are polysemous, did not show up in our dataset.)

Patterns: For all our target–response pairs, we extracted the lexico-syntactic patterns between the targets and the responses. The basic patterns (to be refined; see below) relied on raw frequencies of lemmatised patterns. Since hypernymy requires the definition of pattern direction, all our patterns were marked by their directionality. As corpus resource, we relied on WebKo, a predecessor version of the sdeWaC (cf. Section 3), which comprises more data (approx. 1.5 billion words in comparison to 880 million words) but is less clean. We found a total of 95,615/54,911/21,350 pattern types for the nouns/verbs/adjectives, when neither the length of the patterns was restricted or any kind of generalisation applied. The basic patterns were varied as follows.

---

[4] Ambiguity between the relations arose when the same response was provided for a target with regard to two semantic relations. For example, *Maschine* 'machine' was provided both as a synonym and a hypernym of the noun *Gerät* 'device, machine'. We disregarded such ambiguous cases in this paper.

1. *Morpho-Syntactic Generalisation:* The patterns were generalised by (i) substituting each common noun, proper name, adjective and determiner by its part-of-speech; (ii) deleting all non-alphabetic characters from the patterns.

2. *Mutual Information Variants:* We used point-wise mutual information values (pmi) [51, 18] instead of raw pattern frequencies, and implemented two variants: (i) *pmi(relation,pattern)* and (ii) *pmi(pair,pattern)*, thus enforcing the strengths of patterns that were (i) strong indicators of a specific relation or (ii) strong indicators for specific pairs.

3. *Length Restriction:* The lengths of the patterns were restricted to maximally 1, 2, ... 100 words between the targets and the responses.

4. *Frequency Restriction:* Only patterns with a frequency of at least 1, 2, ... 10, 20, 50, 100 were taken into account, ignoring low-frequent patterns.

5. *Reliability:* The least reliable patterns were deleted from the vector space dimensions. Reliability was determined as in [18]:

$$reliability(pattern) = \frac{\sum_{i \in I} \left(\frac{pmi(i,pattern)}{max_{pmi}} \right) \times reliability_i(i)}{|I|} \tag{1}$$

with i representing a pair instance and I the set of all pairs. The value of $reliability_i$ was instantiated by the strength of the pair in our dataset.

Classification and Evaluation: We implemented a simple[5] *nearest-centroid classification* (also known as *Rocchio Classifier* [52]) to distinguish between the paradigmatic relation pairs. For each word class, we calculated three mean vectors, one for each lexical semantic relation (antonymy, synonymy, hypernymy), as based on the training pairs. We then predicted the semantic relation for the test pairs in each word class, by choosing for each test pair the most similar mean vector, as determined by *cosine*.

This 3-way classification to distinguish between the three paradigmatic relations was performed across the various conditions described above, to identify the types and variations of patterns that were most useful. In a follow-up step we applied the most successful condition to 2-way classifications that aimed to distinguish between two paradigmatic relations (antonyms vs. synonyms, antonyms vs. hypernyms, synonyms vs. hypernyms). The 2-way classifications were to provide insight into more or less difficult relation pairings.

All predictions were evaluated by *precision*, the proportion of predictions we made that were correct. Since many variations of the pattern features effected the number of patterns, we also calculated *recall*, the proportion of test pairs for which we could make a prediction based on the vector dimensions. Harmonic *f-score* then helped us to decide about the overall quality of the conditions in relation to each other.

---

[5] We also applied standard approaches that were relevant to the task, such as Decision Trees and k-Nearest-Neighbour, but our simple approach outperformed them.

4.2   Results

Table 3 shows the results of the pattern-based distinctions in the 3-way relation classification experiments. In the first column the result relies on the basic setup, i.e., using all unaltered patterns as vector features. This *basic* result outperforms the majority baseline (44%) significantly[6] ($p < 0.05$), and is at the same time (a) significantly better than relying on the part-of-speech generalisation ($p < 0.1$), (b) not significantly better than relying on the alphanumeric generalisation and (c) significantly better than the pmi versions of the patterns. Interestingly, optimising the patterns by disregarding very long patterns or disregarding patterns with low frequencies does not improve the basic setup: the best results of these optimisations (see columns *length* and *freq* in Table 3) are exactly the same.

Applying the basic setup to monosemous (*mono*) vs. polysemous (*poly*) relation pairs demonstrates that (a) the lexical semantic relations for monosemous word pairs are easier to predict than for pairs involving polysemy (precision: 64.71 vs. 53.01); (b) the polysemous word pairs activate more pattern types (recall: 45.83 vs. 36.67). Both *mono* and *poly* are significantly better than the baseline ($p < 0.1$).

Table 3. 3-way classification results across conditions

| | basic | generalisations | | length | freq | pmi | | mono | poly |
		pos	alpha			rel,pat	pair,pat		
precision	**59.80	46.85	52.94	**59.80	**59.80	48.04	35.29	*64.71	*53.01
recall	48.41	41.27	42.86	48.41	48.41	38.89	28.57	36.67	45.83
f-score	53.51	43.88	47.37	53.51	53.51	42.98	31.58	46.81	49.16

Figures 1 to 3 show the impact of reducing the number and types of patterns with regard to length, frequency and reliability. Figure 1 demonstrates that reducing the vector space to short patterns of maximally 1, 2, ..., 10 words (i.e., deleting very long and specific patterns that appeared between targets and responses) does almost have no impact on the prediction results. In fact, all patterns seem to provide salient information for the classification, as precision, recall and f-score all monotonically increase when including more and longer patterns. The difference between the best result (including all patterns) and the worst result (including only patterns of length 1) is however not significant. Figure 2 demonstrates that deleting infrequent patterns from the vector space (i.e., deleting patterns with a frequency of less than 2, 3, ..., 100) does also have no strong impact on the prediction results. Even low-frequent patterns seem to provide salient information for the classification, as precision, recall and f-score all monotonically decrease when including only more frequent patterns. Again, the difference between the best result (including all patterns) and the worst result (including only high-frequent patterns) is not significant. Figure 3 shows the effect of deleting x% of the most unreliable patterns, with x = 0.01%, 0.02%,

---

[6] All significance tests have been performed with χ^2. Significance levels are marked at the precision values with $*p \leq 0.1$, $**p \leq 0.05$ and $***p \leq 0.01$, if applicable.

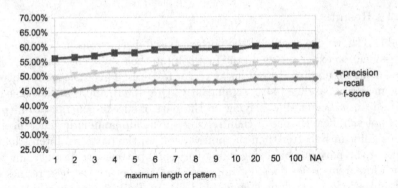

Fig. 1. Deleting long patterns

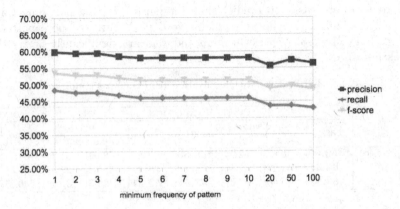

Fig. 2. Deleting infrequent patterns

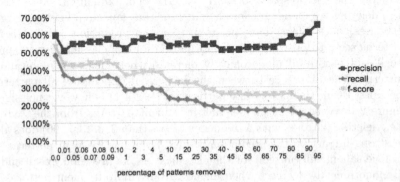

Fig. 3. Deleting unreliable patterns

..., 1%, 2%, ..., 10%, 15%, 20%, ..., 95%. The plot demonstrates that deleting unreliable patterns does have an impact on the quality of the prediction. Most notably, precision drops severely from 59.80% to 51.09% when deleting the most unreliable patterns, and goes up to 65% when only using the 5-10% most reliable patterns. Recall and f-score monotonically decrease when deleting patterns. Even unreliable patterns seem to provide salient information for the classification. At the same time, we achieved our best precision with 5-10% of the patterns only.

Table 4 shows the results of the pattern-based distinctions in the 2-way relation classification experiments. As the baselines are different,[7] we list them in the table. The table demonstrates that the pair-wise distinction between the relation pairs works differently well for the three types. The antonym/synonym distinction performed best, the synonym/hypernym distinction performed worst. While both the antonym/synonym and the antonym/hypernym distinction are significantly better than the baseline, the synonym/hypernym distinction is not.

Table 4. 2-way classification results

	ANT/SYN	ANT/HYP	SYN/HYP
baseline	55.00	50.00	55.00
precision	***78.79	**68.06	63.64
recall	64.20	55.68	50.60
f-score	70.75	61.25	56.38

Table 5 shows the confusion matrix for the 2-way relation distinctions, along with the respective precision scores. The *all* column corresponds to the results in Table 4, the other columns distribute these counts over the word classes. Across the three word classes (*all*), the distinction between antonyms and synonyms is significantly better ($p < 0.1$) than the distinction between synonyms and hypernyms. The other differences (ANT/SYN vs. ANT/HYP; ANT/HYP vs. SYN/HYP) are not significant. So the most difficult relation distinction to predict is synonyms vs. hypernyms.

The confusion matrix demonstrates where the incorrect predictions mainly come from: in the antonym/hypernym distinction, half of the hypernyms were predicted as antonyms; in the synonym/hypernym distinction, even more than half of the hypernyms were predicted as synonyms. While there are also other incorrect predictions, these two cases are striking.

Looking at the results with regard to the three word classes, the predictions of verb relations were in all 2-way distinctions worse than those for nouns and adjectives. The differences for verbs vs. nouns on predicting the synonym/hypernym distinction is significant ($p < 0.05$), the other differences are not significant. The noun and adjective relation prediction is similarly good, without remarkable differences, even though one might have expected that the predictions of the 'core' relations (synonymy and hypernymy for nouns; synonymy and antonymy for adjectives) should be better with regard to the respective word class.

---

[7] Since there are different amounts of antonym/synonym, antonym/hypernym and synonym/hypernym pairs in the final dataset, the majority baseline varies.

Table 5. Confusion matrix (2-way relation distinction)

	NOUN			VERB			ADJ			all		
	ANT	SYN	prec	ANT	SYN	prec	ANT	SYN	prec	ANT	SYN	prec
ANT	**15**	2	77.42	**7**	3	70.59	**8**	1	88.89	**30**	6	78.79
SYN	5	**9**		2	**5**		1	**8**		8	**22**	
	ANT	HYP	prec	ANT	HYP	prec	ANT	HYP	prec	ANT	HYP	prec
ANT	**15**	2	74.19	**8**	2	54.55	**8**	1	73.68	**31**	5	68.06
HYP	6	**8**		8	**4**		4	**6**		18	**18**	
	SYN	HYP	prec	SYN	HYP	prec	SYN	HYP	prec	SYN	HYP	prec
SYN	**13**	1	75.00	**5**	2	42.11	**8**	1	68.42	**26**	4	63.64
HYP	6	**8**		9	**3**		5	**5**		20	**16**	

5 Discussion

The results in the previous section demonstrated that a pattern-based vector space model is able to distinguish between paradigmatic relations: The precision of our basic pattern set in the 3-way relation classification (59.80%) significantly outperformed the majority baseline, $p<0.05$. In the 2-way relation classification, the same patterns achieved precision values of 78.79% for antonym/synonym distinction (significant, $p<0.01$), 68.06% for antonym/hypernym distinction (significant, $p<0.05$), and 63.64% for synonym/hypernym distinction (not significant).

None of the variations to the patterns we performed resulted in significant improvements of the basic setup. Even more, generalisations of the patterns by (i) replacing words with their parts-of-speech or by (ii) deleting all non-alphabetic characters made the results worse, in case (i) even significantly ($p < 0.1$). Similarly, the precision results decreased (in some cases even significantly) when we applied mathematical variations and filters to the patterns, by (i) replacing the pattern frequencies by point-wise mutual information scores as well as when (ii) incorporating a filter for unreliable patterns as adopted from [18].

On the one hand, it is not surprising that generalisations of patterns are not successful because it is very difficult to identify –within a large-scale vector space– those aspects of patterns that contribute to subtle distinctions between relation pairs, and those that will not. For example, if we generalise over specific words by their parts-of-speech this might be helpful in some cases (e.g., we find 'und zwei' *(and two)* as well as 'und sieben' *(and seven)*, where we could generalise over the cardinal number) but contra-productive in others (e.g., we find 'Haar und' *(hair and)* as strong indicator for adjective antonyms and 'Land und' *(country and)* as strong indicator for adjective hypernyms, where generalising over nouns would delete the relation-specific distinction). Similarly, generalising over punctuation might be helpful in some cases (e.g., we find 'und d Arme immer' (lemmatised version of 'und die Armen immer') *and the poor always* as well as ', d Arme immer' , *the poor always* as strong indicators for adjective antonyms) but contra-productive in others (e.g., '/' is a strong indicator for adjective antonymy, while '(' is a strong indicator for adjective hypernymy, and ',' is a strong indicator across all adjectival relations).

On the other hand, we would have expected pmi variants to have a positive effect on the prediction strength of the patterns because they should be able to strengthen the contributions of more salient and weaken the contributions of less salient patterns or pairs. Of course, it is possible that our experimental setup does not sufficiently enforce strong features to outplay weak features. In previous work, many of the existing approaches [8, 16, 15, 18] worked within a bootstrapping cycle, i.e., (1) starting with a small set of clearly distinguishing patterns for a small set of prototypical relation instances, (2) increasing the set of relation pairs on the basis of these patterns and large-scale corpus data, (3) using the new pairs to identify new patterns, (4) filtering the patterns for reliability, etc. It was beyond the scope of this study but might be interesting to implement a variant of our setup that incorporates a bootstrapping cycle. However, we would like to emphasise that we doubt that bootstrapping improves our results because our experiments clearly demonstrated that the salient information in the patterns lies within infrequent as well as frequent patterns, and within short as well as long patterns, and within less reliable as well as strongly reliable patterns. This is in accordance with [19] who demonstrated that large-scale and potentially noisy patterns outperform feature vectors with carefully chosen patterns.

It is difficult to numerically compare our results with related work on pattern-based relations because (i) many previous approaches have tried to *identify* semantic relations pairs, rather than distinguish them [8, 16, 17, 15, 18], and (ii) most of the approaches focused on one semantic relation at the same time [8, 16–18]. Concerning (i), our approach is different in that we *distinguish* between relation pairs; we could however also apply our classification to *identify* additional relation pairs, assuming that we first extract a set of candidate pairs. Concerning (ii), our approach is different in that we focus on 2 or 3 semantic relations at the same time, and in addition the distributional differences between paradigmatic relations are subtle (cf. Section 1). With regard to both (i) and (ii), Turney's work is most similar to ours. [11] achieved a precision of 75% on a set of 136 synonym/antonym questions, with a majority class baseline of 65.4%, in comparison to our synonym/antonym distinction achieving 78.79% with a majority baseline of 55%.

6   Conclusion

This paper presented a vector space model relying on lexico-syntactic patterns to distinguish between the paradigmatic relations *synonymy, antonymy* and *hypernymy*. Our best results achieved a precision score of 59.80%, which significantly outperformed the majority baseline. Interestingly, our original noisy patterns performed better than any kind of standard generalisation or reliability filter. We also showed that (i) antonym/synonym distinction is performed significantly better than synonym/hypernym distinction; (ii) paradigmatic relations between verbs are more difficult to predict than paradigmatic relations between nouns or adjectives; and (iii) paradigmatic relations between monosemous words are easier to predict than those involving a polysemous word.

References

1. Murphy, M.L.: Semantic Relations and the Lexicon. Cambridge University Press (2003)
2. Edmonds, P., Hirst, G.: Near-Synonymy and Lexical Choice. Computational Linguistics 28(2), 105–144 (2002)
3. Curran, J.: From Distributional to Semantic Similarity. PhD thesis, Institute for Communicating and Collaborative Systems, School of Informatics. University of Edinburgh (2003)
4. van der Plas, L., Tiedemann, J.: Finding Synonyms using Automatic Word Alignment and Measures of Distributional Similarity. In: Proceedings of the COLING/ACL 2006 Main Conference Poster Sessions, Sydney, Australia, pp. 866–873 (2006)
5. Fellbaum, C.: Co-Occurrence and Antonymy. Lexicography 8(4), 281–303 (1995)
6. Harabagiu, S.M., Hickl, A., Lacatusu, F.: Negation, Contrast and Contradiction in Text Processing. In: Proceedings of the 21st National Conference on Artificial Intelligence, Boston, MA, pp. 755–762 (2006)
7. Mohammad, S., Dorr, B., Hirst, G.: Computing Word-Pair Antonymy. In: Proceedings of the Conference on Empirical Methods in Natural Language Processing and Computational Natural Language Learning, Waikiki, Hawaii, pp. 982–991 (2008)
8. Hearst, M.: Automatic Acquisition of Hyponyms from Large Text Corpora. In: Proceedings of the 14th International Conference on Computational Linguistics, Nantes, France, pp. 539–545 (1992)
9. Caraballo, S.A.: Automatic Acquisition of a Hypernym-labeled Noun Hierarchy from Text. PhD thesis, Brown University (2001)
10. Snow, R., Jurafsky, D., Ng, A.Y.: Learning Syntactic Patterns for Automatic Hypernym Discovery. Advances in Neural Information Processing Systems 17, 1297–1304 (2004)
11. Turney, P.D.: A Uniform Approach to Analogies, Synonyms, Antonyms, and Associations. In: Proceedings of the 22nd International Conference on Computational Linguistics, Manchester, UK, pp. 905–912 (2008)
12. Yih, W.T., Zweig, G., Platt, J.C.: Polarity Inducing Latent Semantic Analysis. In: Proceedings of the Joint Conference on Empirical Methods in Natural Language Processing and Computational Natural Language Learning, Jeju Island, Korea, pp. 1212–1222 (2012)
13. Mohammad, S.M., Dorr, B.J., Hirst, G., Turney, P.D.: Computing Lexical Contrast. Computational Linguistics 39(3) (to appear, 2013)
14. Berland, M., Charniak, E.: Finding Parts in Very Large Corpora. In: Proceedings of the 37th Annual Meeting of the Association for Computational Linguistics, Maryland, MD, pp. 57–64 (1999)
15. Girju, R., Badulescu, A., Moldovan, D.: Automatic Discovery of Part-Whole Relations. Computational Linguistics 32(1), 83–135 (2006)
16. Girju, R.: Automatic Detection of Causal Relations for Question Answering. In: Proceedings of the ACL Workshop on Multilingual Summarization and Question Answering – Machine Learning and Beyond, Sapporo, Japan, pp. 76–83 (2003)
17. Chklovski, T., Pantel, P.: VerbOcean: Mining the Web for Fine-Grained Semantic Verb Relations. In: Proceedings of the Conference on Empirical Methods in Natural Language Processing, Barcelona, Spain, pp. 33–40 (2004)

18. Pantel, P., Pennacchiotti, M.: Espresso: Leveraging Generic Patterns for Automatically Harvesting Semantic Relations. In: Proceedings of the 21st International Conference on Computational Linguistics and the 44th Annual Meeting of the Association for Computational Linguistics, Sydney, Australia, pp. 113–120 (2006)
19. Turney, P.D.: Similarity of Semantic Relations. Computational Linguistics 32(3), 379–416 (2006)
20. Edmonds, P.: Choosing the Word most typical in Context using a Lexical Co-occurrence Netword. In: Proceedings of the 35th Annual Meeting of the Association for Computational Linguistics, Madrid, Spain, pp. 507–509 (1997)
21. Edmonds, P.: Translating Near-Synonyms: Possibilities and Preferences in the Interlingua. In: Proceedings of the AMTA/SIG-IL Second Workshop on Interlinguas, Langhorne, PA, pp. 23–30 (1998)
22. Edmonds, P.: Semantic Representations of Near-Synonyms for Automatic Lexical Choice. PhD thesis, Department of Computer Science. University of Toronto, Published as technical report CSRI-399 (1999)
23. Curran, J.: Ensemble Methods for Automatic Thesaurus Extraction. In: Proceedings of the Conference on Empirical Methods in Natural Language Processing, pp. 222–229 (2002)
24. Lin, D., Zhao, S., Qin, L., Zhou, M.: Identifying Synonyms among Distributionally Similar Words. In: Proceedings of the International Conferences on Artificial Intelligence, Acapulco, Mexico, pp. 1492–1493 (2003)
25. Erk, K., Padó, S.: A Structured Vector Space Model for Word Meaning in Context. In: Proceedings of the Conference on Empirical Methods in Natural Language Processing and Computational Natural Language Learning, Waikiki, Hawaii, pp. 897–906 (2008)
26. Weeds, J., Weir, D., McCarthy, D.: Characterising Measures of Lexical Distributional Similarity. In: Proceedings of the 20th International Conference of Computational Linguistics, Geneva, Switzerland, pp. 1015–1021 (2004)
27. Lenci, A., Benotto, G.: Identifying Hypernyms in Distributional Semantic Spaces. In: Proceedings of the 1st Joint Conference on Lexical and Computational Semantics, Montréal, Canada, pp. 75–79 (2012)
28. Caraballo, S.A.: Automatic Construction of a Hypernym-labeled Noun Hierarchy from Text. In: Proceedings of the 37th Annual Meeting of the Association for Computational Linguistics, Maryland, MD, pp. 120–126 (1999)
29. Velardi, P., Fabriani, P., Missikoff, M.: Using Text Processing Techniques to Automatically enrich a Domain Ontology. In: Proceedings of the International Conference on Formal Ontology in Information Systems, Ogunquit, ME, pp. 270–284 (2001)
30. Cimiano, P., Schmidt-Thieme, L., Pivk, A., Staab, S.: Learning Taxonomic Relations from Heterogeneous Evidence. In: Proceedings of the ECAI Workshop on Ontology Learning and Population (2004)
31. Snow, R., Jurafsky, D., Ng, A.Y.: Semantic Taxonomy Induction from Heterogenous Evidence. In: Proceedings of the 45th Annual Meeting of the Association for Computational Linguistics, Sydney, Australia, pp. 801–808 (2006)
32. Fellbaum, C.: English Verbs as a Semantic Net. Journal of Lexicography 3(4), 278–301 (1990)
33. Fellbaum, C., Chaffin, R.: Some Principles of the Organization of Verbs in the Mental Lexicon. In: Proceedings of the 12th Annual Conference of the Cognitive Science Society of America, pp. 420–427 (1990)
34. Fellbaum, C.: A Semantic Network of English Verbs. In: [46], pp. 69–104

35. Charles, W., Miller, G.: Contexts of Antonymous Adjectives. Applied Psycholinguistics 10, 357–375 (1989)
36. Justeson, J.S., Katz, S.M.: Co-Occurrence of Antonymous Adjectives and their Contexts. Computational Linguistics 17, 1–19 (1991)
37. Lucerto, C., Pinto, D., Jiménez-Salazar, H.: An Automatic Method to Identify Antonymy Relations. In: Proceedings of the IBERAMIA Workshop on Lexical Resources and the Web for Word Sense Disambiguation, Puebla, Mexico, pp. 105–111 (2004)
38. de Marneffe, M.C., Rafferty, A.N., Manning, C.D.: Finding Contradictions in Text. In: Proceedings of the 46th Annual Meeting of the Association for Computational Linguistics: Human Language Technologies, Columbus, OH, pp. 1039–1047 (2008)
39. Lin, D., Pantel, P.: DIRT – Discovery of Inference Rules from Text. In: Proceedings of the ACM Conference on Knowledge Discovery and Data Mining, San Francisco, CA, pp. 323–328 (2001)
40. Turney, P.D.: Measuring Semantic Similarity by Latent Relational Analysis. In: Proceedings of the 19th International Joint Conference on Artificial Intelligence, Edinburgh, Scotland, pp. 1136–1141 (2005)
41. Turney, P.D.: Expressing Implicit Semantic Relations without Supervision. In: Proceedings of the 21st International Conference on Computational Linguistics and the 44th Annual Meeting of the Association for Computational Linguistics, Sydney, Australia, pp. 313–320 (2006)
42. Lin, D.: Automatic Retrieval and Clustering of Similar Words. In: Proceedings of the 17th International Conference on Computational Linguistics, Montreal, Canada, pp. 768–774 (1998)
43. Hamp, B., Feldweg, H.: GermaNet – a Lexical-Semantic Net for German. In: Proceedings of the ACL Workshop on Automatic Information Extraction and Building Lexical Semantic Resources for NLP Applications, Madrid, Spain, pp. 9–15 (1997)
44. Kunze, C.: Extension and Use of GermaNet, a Lexical-Semantic Database. In: Proceedings of the 2nd International Conference on Language Resources and Evaluation, Athens, Greece, pp. 999–1002 (2000)
45. Lemnitzer, L., Kunze, C.: Computerlexikographie. Gunter Narr Verlag, Tübingen (2007)
46. Fellbaum, C. (ed.): WordNet – An Electronic Lexical Database. Language, Speech, and Communication. MIT Press, Cambridge (1998)
47. Faaß, G., Heid, U., Schmid, H.: Design and Application of a Gold Standard for Morphological Analysis: SMOR in Validation. In: Proceedings of the 7th International Conference on Language Resources and Evaluation, Valletta, Malta, pp. 803–810 (2010)
48. Baroni, M., Bernardini, S., Ferraresi, A., Zanchetta, E.: The WaCky Wide Web: A Collection of Very Large Linguistically Processed Web-Crawled Corpora. Language Resources and Evaluation 43(3), 209–226 (2009)
49. Schiller, A., Teufel, S., Stöckert, C., Thielen, C.: Guidelines für das Tagging deutscher Textcorpora mit STTS. Institut für Maschinelle Sprachverarbeitung, Universität Stuttgart, and Seminar für Sprachwissenschaft, Universität Tübingen (1999)
50. Miller, G.A., Fellbaum, C.: Semantic Networks of English. Cognition 41, 197–229 (1991)
51. Church, K.W., Hanks, P.: Word Association Norms, Mutual Information, and Lexicography. Computational Linguistics 16(1), 22–29 (1990)
52. Christopher, D., Manning, P.R., Schütze, H.: Introduction to Information Retrieval. Cambridge University Press (2008)

Dependency-Based Algorithms for Answer Validation Task in Russian Question Answering

Alexander Solovyev

Bauman Moscow State Technical University
asolovyev@lib.bmstu.ru

Abstract. This paper discusses the Answer Validation Task in Question Answering applied for Russian language. Due to poor language resources we are limited in selection of techniques for Question Answering. Dependency parse-based methods applied to factoid questions are in the primary focus. We notice that existing works use either pure syntactic dependency parsers or parsers which perform some extra shallow semantic analysis for English. The selection of either of the parsers is not justified in any of these works. We report experiments for Russian language in absence of WordNet on various combinations of rule-based parsers and graph matching algorithms, including our Parallel Graphs Traversal algorithm first published in ROMIP 2010 [23]. Performance is evaluated using a subset of ROMIP questions collection with ten-fold cross-validation.

Keywords: Information retrieval, question answering, answer validation.

1   Introduction

Question answering is an information retrieval task. Given a question formulated as a natural language statement a system searches for a concise answer within a collection of unstructured texts. It is different from classical information retrieval task, which is to find a set of relevant documents. Various techniques address different kinds of questions: factoid (who? where? when?), definitional (What is ABC?), and more complex: How? and Why? Factoid and definitional questions are known to be significantly easier to deal with automatically [3].

Answer search task can be decomposed into 4 subtasks: Question Analysis, Search, Answer Extraction, and Answer Validation. Fig. 1 illustrates the same decomposition applied to the QA system developed in this work. The Question analysis task is to derive a semantic class of answer and a question focus. Focus is a part of a question that identifies informative expectations expressed by question [1][3]. Examples of foci are: "Which city", "What year", "When", "Who", "How far", "How tall". The most popular semantic classes of answers for factoid questions are: PERSON, LOCATION, TIME, LENGTH, ORGANIZATION, and etc [14].

Non-focus words of a question statement are called question support. These words constitute an important input for Search task: to find statements which contain an answer to the question in a whole full-text collection. Question words are not very useful for searching, because declarative statements are unlikely to contain them.

I. Gurevych, C. Biemann, and T. Zesch (Eds.): GSCL 2013, LNAI 8105, pp. 199–212, 2013.
© Springer-Verlag Berlin Heidelberg 2013

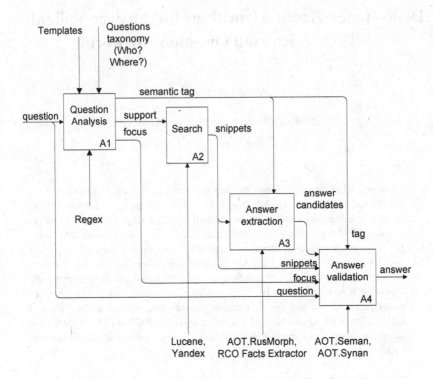

Fig. 1. Functional diagram of generic subtasks in Question answering

The simplest technique for question analysis is to use manually pre-defined regular expressions. A more advanced approach is to train a statistical classifier [6].

A trivial implementation of search task is a straightforward querying of an existing search engine (e.g. *Lucene, Indri, Yandex*) using question support as keywords. Engines usually provide snippets, which are good passage candidates for answer extraction. However, a generic snippet generation method (e.g. *SimpleFragmenter* in *Lucene*) might be not very efficient for question answering, therefore customized passage extraction techniques were developed [14].

The answer extraction task is to derive concise answers from snippets. Two general approaches are used: extracting answers as a sentence and extracting words of a given semantic class. The former uses text summarization techniques [6], the later uses tools for named entities extraction [16][22]. Statistical methods are mostly used for both of them; however rule-based tools are still dominating for Russian language [25].

Early systems used no further processing steps once an answer was extracted. The fourth processing step – Answer Validation – was inspired by another generic NLP task: *Recognizing Textual Entailment (RTE)*. Given a hypothesis statement and a supporting text, it is required to recognize whether this hypothesis follows from the text or not. In case of Question Answering a hypothesis is a combination of a question and a candidate answer, and supporting text is a snippet, a candidate passage, or even a whole document. Many RTE methods involve text parsing which can be performed by many tools. Some tools perform a syntactic parse; others do an extra shallow semantic processing of text.

The list of QA techniques is not limited by RTE, but most of them require rich language resources, which are not available for Russian: there is no satisfactory Word-Net-style ontology, and creating a big annotated textual corpora is still an ongoing project[1,2]. Therefore, we focus on Answer Validation techniques which are applicable without rich ontology and statistical learning on big corpora – Dependency-based RTE. In previous works we developed Parallel Graph Traverse algorithm [23] and measured its performance as part of our QA System Umba on ROMIP 2010 campaign with a single competitor who was not used answer validation at all. This work focuses on answer validation and addresses the following questions:

- How do existing Answer Validation algorithms perform on Russian language using only the rule-based parsers?
- What is the performance of our Parallel Graph Traverse algorithm compared to other known algorithms?
- What is the performance gain of shallow semantic parse use in Answer Validation compared to syntactic dependency parse in absence of WordNet?

2   Related Work

Textual Entailment Recognition used to be an annual task in Text Analysis Conference up to 2011 [2]. There was a subset of tasks identified as QA domain in these competitions. In parallel a series of annual events *Answer Validation Exercise (AVE)* were held as part of CLEF campaign until 2008 [19]. Textual entailment is not the only approach for Answer Validation. Other approaches include projection [16] and exploiting web redundancy [10].

Four general approaches to a textual entailment can be identified: *Lexical overlap*, *Dependency overlap*, *Constituency overlap* [8], and *Logic reasoning* [1][13]. Lexical overlap approach models text as a set of independent unconnected terms (as in Bag-of-words model) and employs a lexical ontology (e.g. WordNet) to weight these terms differently (unlike TF-IDF weighting in traditional IR). Lexical overlap is often used as a backup strategy when syntactic parsing fails [26], however experiments with lexical overlap being the only QA strategy used are reported as well [9]. *Constituency overlap* is not used for Answer Validation, presumably because questions are too short.

Dependency overlap methods can be classified by generic trees matching algorithms: calculating Tree-Edit Distance [18][21], Tree Alignment [5][11], and Maximal Embedded Sub-tree search [4][7][12]. A Predicate Matching method proposed in [18] can be considered as a hybrid of Trees Alignment and Lexical overlap. Dependency-based Answer Validation implementations can also be classified by usage of parsers: syntactic only or shallow-semantic. Having such a big variety of methods for answer validation, some work was done on combining them [15][22].

Table 1 summarizes RTE algorithms and text models used for Answer Validation.

---

[1] http://ruscorpora.ru/en/ Syntactic labels assigned semi-automatically for 539 docs, 49423 sentences, 757794 words. Semantic labels automatically assigned without moderation.
[2] http://opencorpora.org/ (in Russian) – 9% out of 1.6M tokens annotated, 1st June 2013.

This paper is a first work reporting about experiments on Answer Validation for Russian language. Various dependency-based algorithms were applied to syntactic and shallow semantic sentences parsing, including implementation of our own algorithm – Parallel Graph Traversal, which falls into Embedded Sub-tree RTE class. Some of these experiments (A, B, C, D) were not reported before for English as well.

Table 1. Algorithms and text models used in RTE for QA

	Bag-of-words	Syntactic dependencies	Shallow semantics	Logical forms
Sets overlap	[9][26]	A	[26]	
Predicate matching		B	[22]	
Trees Alignment	N/A	[5][11]	C	N/A
Tree-Edit Distance		[18][21]	D	
Embedded sub-trees		[4][7][12]	E	
Automated theorem proving		N/A		[1][13]

Position E corresponds to an algorithm we used in our QA system which was evaluated in ROMIP 2010 campaign on a regular QA track [23][24]. However, in this paper we measure its performance on Answer Validation Exercise and compare it with other algorithms. For each row in Table 1 a single generic algorithm implementation was used, and for every column a single Russian parsing tool was used – either AOT.Seman or AOT.Synan[3]. Performance of different combinations of algorithms and parsers was measured on ROMIP questions collection and manually assessed answers collection based on Yandex snippets.

3   Dependency-Based Answer Validation Methods

In this section we enlist text models and algorithms for comparing dependency parses of questions and snippets, which are used in Answer Validation. For any implementation of Answer Validation module we assume a common interface contract: it assigns a confidence value to tuple <Question, Answer, Snippet>. Some threshold value is then used to make a binary decision – whether the answer is valid or not. Therefore, every method described in this section should end up with a normalized confidence value 0..1.

3.1   Lexical and Triplets Overlap

Given a set of Question terms Q and a set of Snippet terms T we can calculate the percentage of common terms in Question and Snippet:

$$E = |Q \cap T| / |Q| \tag{1}$$

---

[3] http://aot.ru/docs/sokirko/sokirko-candid-eng.html: A short description of Dialing Project. Alexey Sokirko, 2001.

It is also natural to consider that focus words in Question correspond to answer words in Snippet. The same formula can be applied to Triplets <N1,R,N2> where N1 and N2 are nodes in dependency parsing tree and R is the label of dependency.

3.2   Tree Edit Distance

Given the cost of three Tree edit operations [18]: node change, delete and replace $\gamma(a{\rightarrow}\lambda)$, $\gamma(\lambda{\rightarrow}a)$, $\gamma(a{\rightarrow}b)$, the task is to find a sequence S of such operations which will transform a Snippet tree into a Question tree. Out of all possible sequences the algorithm is looking for those which have the minimal total cost. An obvious pre-processing step is to replace Focus terms by *ANS* in a question tree and answer terms by *ANS* placeholder in a snippet to allow for matching.

Considering an observation that a snippet text can contain a lot of additional information, we also would like to allow for removal of sub-trees from a snippet without a penalty. An efficient dynamic programming algorithm for this task was suggested in [27]. Its significant limitation is that it works on ordered trees only; therefore Tree-Edit distance method depends on words order.

In our implementation we transform an edit cost to a confidence score as follows:

$$s = 1 - cost / num\_of\_terms\_in\_question$$

3.3   Trees Alignment

The algorithm is originated from a statistical translation task, which requires an analysis of parallel texts. The task is to build a matrix $M[Nq,Ns]$ with scores of correspondence between terms in a question and terms in a snippet. A score of nodes match is a linear combination of lexical similarity of these two words and the score of all direct child matches. To calculate the match between x child of a given Question node and y child of a given Snippet node all possible combinations between x and y children are considered. The best total score of a combination will indicate the best child match. This value will be used for original nodes score calculation. A parent weight parameter controls weights between lexical similarity of nodes themselves and score of their child match. Another important feature is that the algorithm allows for skipping nodes in snippet by applying skip penalty during matching.

Dynamic programming algorithm for this task is described in [11]. As a score for the whole trees match we chose the matching score between root nodes, which are usually predicates.

3.4   Predicates Matching

In the OpenEphyra system a hybrid algorithm of Lexical Overlap and Trees Alignment is used [22]. It considers a Semantic Role Labeling done by the ASSERT parser [17], which marks predicates and their arguments. The matching of predicates is calculated as a product of lexical similarity of verbs and lexical overlap of their arguments. Lexical overlap is calculated by using a modification of the formula (1), which allows for a fuzzy matching between sets. While calculating intersection of powersets, a Lexical similarity of individual words is considered instead of the strict equality.

In case of Russian language there is no SLR-parser available, so in our experiments we used all terms which transitively dependent on predicate instead of argument terms in original formula from [22].

3.5   Answer Validation Using Parallel Graph Traversal Algorithm

The development of this algorithm was started in our previous work [24]. The algorithm searches for a maximal embedded sub-graph of a Question in a Snippet. Unlike in [7] our implementation forcibly starts searching from a pair of seed nodes, which are required to match: the focus node in a question and the answer node in a snippet. We also address an issue raised in [11] of allowing single intermediate nodes to appear in snippet graph during in-depth search of a maximal sub-graph. We do this by pre-processing which adds shortcut edges to detour every node in a snippet. These edges are labeled as *shortcut* and selecting them during in-depth traversal results in a penalty: shortcut_penalty.

A similarity score for graphs generally correspond to the number of nodes in a maximal embedded sub-graph. However, due to shortcut penalties and penalties for an inexact matching of terms (e.g. matching of lemmas only), following recursive formula should be applied to calculate graphs matching score starting from a pair of focus and answer nodes:

$$s(v_q, v_a) = \begin{cases} 1 + sim_{inc}(v_q, v_a) + sim_{out}(v_q, v_a) & if \ sim(v_q, v_a) > 0 \\ 0 & otherwise \end{cases}$$

$$sim_{inc}(v_q, v_a) = \sum_{e_q \in inc(v_q)} \max_{e_p \in inc(v_a)} s(src(e_q), src(e_a))$$

$$sim_{out}(v_q, v_a) = \sum_{e_q \in out(v_q)} \max_{e_p \in out(v_a)} s(trg(e_q), trg(e_a))$$

Where $s(v,v)$ is a score similarity for two nodes, $src(e)$ and $trg(e)$ are source and target nodes of arc e, $sim(v,v)$ is a Lexical similarity between words. In absence of a Wordnet-like ontology we differentiate two cases: the exact words match and the lemmas match, whose scores are defined as parameters for the algorithm. A total score is then normalized by the number of nodes in a question graph.

4   Experiments

4.1   Test Collection

We chose ROMIP collection of Russian questions for performance evaluation. However, to carry out an AVE-like experiment we need a collection of manually assessed tuples: <Question, Answer, Snippet, Document, True/False>. Such a collection can theoretically be derived from relevance tables produced from ROMIP 2010 QA track, which is a regular campaign for evaluating QA systems on Russian language where we started our experiments [23]. However, the quality of output was not very high. Organizers reported two possible reasons for that: either the document collection did not match the questions collection, or the participant systems did not manage to find useful answers within that collection.

For the purpose of our experiments we decided to create our own test collection for Answer Validation. We took ROMIP 2010 questions (those, which were randomly chosen for manual assessment), chose questions on PERSONS, LOCATIONS, and ORGANIZATIONS, and tried to find answers manually in Yandex search results. For each of the selected questions we have found both positive and negative examples of answers on the first page of Yandex results. The manual assessment was made by only one person. Table 2 contains numerical characteristics of the collection.

Table 2. Russian Factoid Answer Validation test collection volume characteristics

Characteristics\ Answer type	Location	Person	Organization	Total
Questions	26	22	6	54
QA pairs	146	105	20	271
True answers	45	41	11	97
False answers	101	64	9	174
Questions without true answers	4	4	1	9

Question-answer collection created for these experiments was published and is available for downloading in AVE-compatible XML format[4].

4.2   Evaluation Metrics

Traditional metrics for Answer Validation and Textual Entailment Recognition tasks are *Accuracy* and *F-measure*. *F-measure* does not depend on the number of true negative outcomes, which is a significant drawback for measuring performance of wrong answers filtering mechanism. E.g. consider a collection of one positive QA pair and 9 negative, i.e. only one pair was assessed as the right answer for the question. A system which has no Answer Validation mechanism yielding all candidate answers as true (1 true positive, 9 false positives) will get $F_{0.5} = 0.122$. Another system which simply rejects everything as false (1 false negative, 9 true negatives) will get $F_{0.5} = 0$. Given a task to filter false positive answers, the first system makes 9 serious mistakes and gets a higher score than the second system. To compensate for it, we derived a new metric called Weighted Error, which gives different weights to false negatives and positives (as F-measure does), however still gives credit for true negatives:

$$E_\alpha = \frac{\frac{\alpha \cdot fp + fn}{\alpha + 1}}{tp + tn + \frac{\alpha \cdot fp + fn}{\alpha + 1}} = \frac{\alpha \cdot fp + fn}{(\alpha + 1) \cdot (tp + tn) + \alpha \cdot fp + fn}$$

This metric awards more those filters that do not yield wrong answers. As a result a trivial filter which blocks all candidate answers will get fewer penalties than a filter which yields all answers. The goal for development of this metric is similar to goal of c@1 [20] – measuring contribution of non-response into performance. Unlike c@1,

---

[4] http://qa.lib.bmstu.ru/rusave2012.zip

which is a per-question measurement of QA system performance, our metric is designed for use in answer validation sub-task only, which is a binary classification problem for a question-answer pair.

4.3   Configurations

This section describes implementations of Answer Validation modules evaluated in this work and defines the names used in Results section. Some of original methods used WordNet-based similarity together with the algorithms described. There is no WordNet for Russian language, so we didn't reproduce original methods precisely.

Fail All. A trivial singular filter which blocks all candidate answers. It gets quite a low penalty because our collection contains a lot of negative examples. In IR terms it gives maximal precision with zero coverage.

Do Nothing. A trivial filter which yields all candidate answers.

Lexical Overlap. Baseline implementation of simple lexical overlap.

Triplets Overlap [with/no] Labels [AOTSeman/AOTSynan]. Calculates the overlap of sets of term-to-term links in question and snippet as Triple Matcher in [26]. Implementation can demand exact match of links labels or not. It also can use either Syntactic parse or Shallow semantics analysis graph. In [26] a *Minipar* parser was used. The only parameter is *threshold*.

Trees Alignment [AOTSeman/AOTSynan]. An algorithm from [11] is applied to either syntactic parse or shallow semantic analysis graph. In the original work *Malt-Parser* was used. The parameters are: *skip penalty, parent weight, threshold*.

Tree Edit Distance [AOTSeman/AOTSynan]. An algorithm from [18] is applied to either syntactic tree or shallow semantic graph. In the original work *Collins parser* was used. The parameters are: *insert_cost, delete_cost, change_cost*, and *threshold*.

Traversal [AOTSeman/AOTSynan]. An algorithm of Parallel Graph Traversal from [24] which represents an approach of Maximal Embedded Sub-tree. It can be applied to either syntactic dependencies tree or to shallow semantic analysis graph. In the original work only AOT.Seman parser was used. Parameters are: *exact_node_match, lemma_match, shortcut_penalty, threshold*.

Predicate Matching [AOT.Seman/AOT.Synan]. Predicate Matching algorithm from [22] applied to either syntactic dependencies parse or to shallow semantic analysis graph. In the original work the *ASSERT* parser was used, but in our experiments we apply AOT.Synan and AOT.Seman parsers. The parameters are: *not_equals_term_similarity, threshold*. The former is a mechanism to simulate the presence of WordNet-based terms similarity, which cannot be calculated for Russian language, however is affects original calculations significantly. The automatic parameters optimization gives low value to this parameter. Forcibly setting it to zero does not change the results much.

4.4   Parameters Tuning

We generally follow the AVE evaluation procedure. In both AVE and RTE campaigns organizers provide participants with so-called *Development collection (or training collection)* – an assessed set of tasks which can be used to tune the system for best performance on some real data.

These collections are usually full output tables of a previous year competition. On some deadline participants receive a second unsupervised collection without assessors marks – just triplets *<question,answer,context>*. This is a *Test collection*, for which a participant system should make a decision whether it is true or false. Organizers publish a gold-standard markup of the Test collection after all participants submit results.

Our first experiment is equivalent to the Train phase of AVE/RTE challenges – we use our collection for development purpose and tune the parameters to achieve best results in every configuration. In section 4.5 we report the results of cross-validation. Table 3 contains the results of parameters optimization for every configuration. A goal for optimization is to minimize the weighted error $E_{2.0}$. An optimization is performed via the brute-force scanning of all parameters with step 0.01 in a range $0...1$. I.e. to evaluate Tree Edit Distance algorithm we'd have to examine $2*100^4$ tuples of parameters (3 parameters, threshold value, and 2 parses – syntactic and semantic). Two techniques were used to reduce the calculation cost:

- All parses were cached. Therefore, the most expensive operation of syntactic parse was performed once for whole collection.
- For every tuple of parameters excluding threshold and every QA pair we calculated a score value. For this set of scores we searched for the best threshold value. Therefore, scanning for best threshold did not involving Answer Validation. I.e. for Tree Edit distance we actually ran every algorithm $2*100^3$ times.

A remarkable result is that trivial *Fail all* run showed a median result: 8 out of 15 runs performed worse, and 6 – better than *Fail all* run. So, in our experiments *Fail all* happens to be a much more reasonable baseline than traditional Lexical overlap.

Another observation which one can make is that there is no simple answer whether to use a shallow semantic analysis or not. Various algorithms behave in different ways and the benefit from switching from syntax to semantics or vice versa can be less than statistical error. At least our collection is too small for measuring this difference for every method. However, two algorithms demonstrate noticeable decrease of quality when adding shallow semantic processing: *Triplets Overlap with Labels* and *Tree Edit Distance*. Behavior of the Triplets overlap algorithm confirms other researchers' observations of uselessness of labels matching in RTE. This can be explained by the large label assignment variability or inaccuracy. That error multiplies when another layer of logic added for assigning other semantic labels above old (inaccurate) syntactic labels. A worse performance of the second one can be explained by the dependence of the tree-edit distance on the words order. This order becomes irrelevant in semantic graph representation and therefore there is no obvious strategy of ordering "semantic trees". In our implementation we used ordering by words position, which is a replica of the original algorithm behavior. To check the statistical significance of these observations we performed a cross-validation over same collection.

Table 3. Performance metrics for various configurations of Answer Validation module. Optimal values are reported (parameters which minimizes $E_{2.0}$).

| Configuration | tp | fp | Accu- | | |
	fn	tn	racy %	$F_{0.5}\%$	$E_{2.0}\%$
Fail all	0	0			
	97	174	64.2	0.0	15.7
Do nothing	97	174			
	0	0	35.8	41.1	54.5
Lexical Overlap (0.9)	49	48			
	55	119	62.0	49.8	23.1
Triplets Overlap no labels (th=0.67) AOT.Seman	17	8			
	89	166	65.4	41.3	16.1
Triplets Overlap with labels (0.875) AOT.Seman	20	17			
	59	130	66.4	44.1	17.1
Triplets Overlap no labels (th=0.76) AOT.Synan	**14**	**2**			
	82	**172**	68.9	43.8	13.4
Triplets Overlap with labels (0.775) AOT.Synan	6	1			
	73	146	67.3	28.0	14.1
Trees alignment (sp=1. pw=0.95. th=0.95) AOT.Seman	43	39			
	53	135	65.9	50.7	19.7
Trees alignment (sp=0.35 pw=0.85 th=0.9) AOT.Synan	38	43			
	58	131	62.6	45.2	22.1
Tree edit distance (insert=1.0 del=0.04 change=0.91 th=0.97) AOT.Seman	9	11			
	87	163	63.7	25.6	17.4
Tree edit distance (insert=1.0 del=0.64 change=0.96 th=0.79) AOT.Synan	10	13			
	86	161	63.3	26.6	17.9
Traversal (match=0.27 lem=0.12 shortcut=0.5 th=0.05) AOT.Seman	**27**	**8**			
	69	**166**	71.5	57.2	12.8
Traversal (match=0.09 lem=0.11 shortcut=0.5 th=0.04) AOT.Synan	**15**	**4**			
	81	**170**	68.5	43.6	13.8
Predicate(ne=0.01 th=0.41) AOT.Synan	7	2			
	89	172	66.3	26.5	14.8
Predicate(ne=0.26 th=0.01) AOT.Seman	**26**	**10**			
	70	**164**	70.4	54.2	13.6

4.5   Cross Validation

Our second experiment is a 10-fold cross-validation of algorithms on the same collection. The collection is randomly divided into 10 subsets. Ten two-phase experiments are then conducted: a development phase and a test phase. For every i^{th} subset we optimize parameters based on the rest 9 subsets and measure the performance on i^{th} subset. After obtaining ten $E_{2.0}$ values for every configuration we calculate average value and standard deviation. Table 4 contains results for every configuration measured for every of 10 folds. Note, unlike Table 3 it does not contains specific optimal parameters values because they are different for every of 10 folds.

Table 4. Results of 10-fold cross-validation

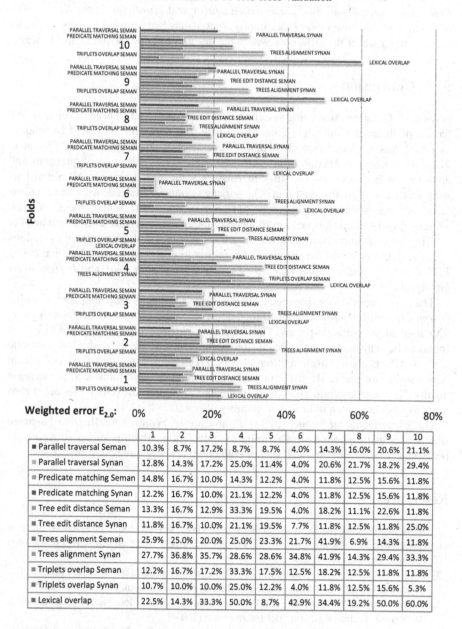

	1	2	3	4	5	6	7	8	9	10
▪ Parallel traversal Seman	10.3%	8.7%	17.2%	8.7%	8.7%	4.0%	14.3%	16.0%	20.6%	21.1%
▪ Parallel traversal Synan	12.8%	14.3%	17.2%	25.0%	11.4%	4.0%	20.6%	21.7%	18.2%	29.4%
▪ Predicate matching Seman	14.8%	16.7%	10.0%	14.3%	12.2%	4.0%	11.8%	12.5%	15.6%	11.8%
▪ Predicate matching Synan	12.2%	16.7%	10.0%	21.1%	12.2%	4.0%	11.8%	12.5%	15.6%	11.8%
▪ Tree edit distance Seman	13.3%	16.7%	12.9%	33.3%	19.5%	4.0%	18.2%	11.1%	22.6%	11.8%
▪ Tree edit distance Synan	11.8%	16.7%	10.0%	21.1%	19.5%	7.7%	11.8%	12.5%	11.8%	25.0%
▪ Trees alignment Seman	25.9%	25.0%	20.0%	25.0%	23.3%	21.7%	41.9%	6.9%	14.3%	11.8%
▪ Trees alignment Synan	27.7%	36.8%	35.7%	28.6%	28.6%	34.8%	41.9%	14.3%	29.4%	33.3%
▪ Triplets overlap Seman	12.2%	16.7%	17.2%	33.3%	17.5%	12.5%	18.2%	12.5%	11.8%	11.8%
▪ Triplets overlap Synan	10.7%	10.0%	10.0%	25.0%	12.2%	4.0%	11.8%	12.5%	15.6%	5.3%
▪ Lexical overlap	22.5%	14.3%	33.3%	50.0%	8.7%	42.9%	34.4%	19.2%	50.0%	60.0%

To tune parameters in reasonable time we scanned them with step=0.1, since cross-validation requires 10 times more computations than the first experiment. This session lasted 3 days on eight cores.

To evaluate statistical significance of advantage of one configuration over another a one-tailed T-test with p=0.05 has been used. A group of leading configurations (having lower weighted error values) has been discovered: *Triplets overlap Synan*, *Predicates match Synan*, *Predicates match Seman*, and *Parallel traversal Seman*. The advantage of this group over other configurations has been proved to be statistically significant; however the difference within the group for our collection is insignificant.

5   Conclusion

A research Question Answering system for Russian language has been developed. We have reproduced existing algorithms to evaluate four major techniques of Dependency-based Answer Validation: *Lexical overlap*, *Triplets Overlap*, *Tree-edit Distance*, *Trees Alignment* and *Maximal Embedded Sub-tree*. A test collection of questions and answers has been created for Russian Answer Validation Exercise. All algorithms were evaluated with two kinds of text parsing: syntactic dependencies and shallow semantic analysis using rule-based AOT.Synan and AOT.Seman tools. A new Weighted Error metric has been introduced to measure the performance of the answer filtering by awarding systems for true negative outcomes and differentiating penalties for false negatives and false positives. According to this new metric a baseline trivial *Fail all* strategy outperforms *Tree Edit Distance*, *Trees alignment*, and *Lexical overlap* algorithms.

The best performance has been achieved by *Predicate Matching* and *Parallel Graph Traversal* algorithms applied to shallow semantic parses and by *Triplets overlaps* applied on syntactic parses. The difference between these algorithms has been found to be statistically insignificant. Replacing syntactic parse by shallow semantic analysis affects algorithms in a different way. *Tree Edit distance* and *Triplets Overlap* demonstrate noticeable decrease in performance; other algorithms tend to increase quality. The lead configurations in both development and test (cross-validated) runs are almost same, which indicates on stability of results.

Our experiments are limited by (a) an absence of Wordnet for Russian language (that forces us to ignore a term's similarity calculation, which is supposed to impact performance a lot); (b) the small size of collection; (c) only three question classes.

References

1. Akhmatova, E., Molla, D.: Recognizing textual entailment via atomic propositions. In: Proceedings of the Machine Learning Challenges Workshop, Southampton, UK, pp. 385–403 (2006)
2. Bentivogli, L., Clark, P., Dagan, I., Dang, H., Giampiccolo, D.: The Seventh PASCAL Recognizing Textual Entailment Challenge. In: TAC 2011 Notebook Proceedings (2011)
3. Burger, J., et al.: Issues. Tasks and Program Structures to Roadmap Research in Question Answering. Technical Report. SRI International (2003)
4. Chen, P., Ding, W., Simmons, T., Lacayo, C.: Parsing Tree Matching Based Question Answering. In: Proc. of the Text Analysis Conference Workshop, Gaithersburg. USA (2008)
5. Li, F., et al.: THU QUANTA at TAC 2008 QA and RTE track. In: Text Analysis Conference (TAC 2008), Gaithersburg, Maryland USA (2008)

6. Ittycheriah, A.: A Statistical Approach For Open Domain Question Answering. In: Advances in Open Domain Question Answering, vol. 32, pp. 35–69. Springer, Netherlands (2006)
7. Katrenko, S., Adriaans, P.: Using Maximal Embedded Syntactic Subtrees for Textual Entailment Recognition. In: Proceedings of RTE-2 Workshop, Venice, Italy (2006)
8. Katrenko, S., Toledo, A.: A Comparison Study of Lexical and Syntactic Overlap for Textual Entailment Recognition. In: Proceedings of BNAIC 2011 (2011)
9. Ligozat, A.-L., Grau, B., Vilnat, A., Robba, I.: Arnaud Grappy: Lexical validation of answers in Question Answering. Web Intelligence 2007, 330–333 (2007)
10. Magnini, B., Negri, M., Prevete, R., Tanev, H.: Is it the right answer?: exploiting web redundancy for Answer Validation. In: Proceedings of the 40th Annual Meeting on Association for Computational Linguistics, Philadelphia, Pennsylvania (2002)
11. Marsi, E.C., Krahmer, E.J., Bosma, W.E., Theune, M.: Normalized Alignment of Dependency Trees for Detecting Textual Entailment. In: Second PASCAL Recognising Textual Entailment Challenge, Venice, Italy, April 10-12, pp. 56–61 (2006)
12. Micol, D., Ferrández, O., Munoz, R., Palomar, M.: DLSITE-2: Semantic similarity based on syntactic dependency trees applied to textual entailment. In: Proceedings of the Text-Graphs-2 Workshop, Rochester, NY, USA, pp. 73–80 (2007)
13. Moldovan, D., Pasca, M., Surdeanu, M.: Some Advanced Features of LCC's PowerAnswer. In: Advances in Open Domain Question Answering, Text, Speech and Language Technology, Part 1, vol. 32, pp. 3–34 (2006)
14. Moldovan, D., et al.: LASSO: a tool for surfing the answer net. In: Proceedings of the 8th TExt Retrieval Conference (TREC-8), Gaithersburg, Maryland (1999)
15. Pakray, P., et al.: A Hybrid Question Answering System based on Information Retrieval and Answer Validation. CLEF (Notebook Papers/Labs/Workshop) (2011)
16. Prager, J.: Open-Domain Question–Answering. In: Foundations and Trends® in Information Retrieval, vol. 1(2), pp. 91–231 (2006), http://dx.doi.org/10.1561/1500000001
17. Pradhan, S., Ward, W., Hacioglu, K., Martin, J., Jurafsky, D.: Shallow Semantic Parsing using Support Vector Machines. In: Proceedings of the Human Language Technology Conference/North American Chapter of the Association for Computational Linguistics Annual Meeting (HLT/NAACL-2004), Boston, MA, May 2-7 (2004)
18. Punyakanok, V., Roth, D., Yih, W.: Mapping Dependencies Trees: An Application to Question Answering. In: Proceedings of AI&Math 2004 (2004)
19. Rodrigo, Á., Peñas, A., Verdejo, F.: Overview of the answer validation exercise 2008. In: Peters, C., Deselaers, T., Ferro, N., Gonzalo, J., Jones, G.J.F., Kurimo, M., Mandl, T., Peñas, A., Petras, V. (eds.) CLEF 2008. LNCS, vol. 5706, pp. 296–313. Springer, Heidelberg (2009)
20. Peñas, A., Rodrigo, A.: A Simple Measure to Assess Non-response. In: Proceedings of 49th Annual Meeting of the Association for Computational Linguistics - Human Language Technologies, ACL-HLT 2011 (2011)
21. Schilder, F., McInnes, B.T.: Word and tree-based similarities for textual entailment. In: Magnini, B., Dagan, I. (eds.) Proceedings of the Second PASCAL Workshop on Textual Entailment, Venice, Italy, pp. 140–145 (2006)
22. Schlaefer, N.: A Semantic Approach to Question Answering (2007)
23. Solovyev, A.: Who is to blame and Where the dog is buried? Method of answers validations based on fuzzy matching of semantic graphs in Question answering system. In: Proceedings of ROMIP 2011, Kazan, pp. 125–141 (2011)

24. Solovyev, A.: Algorithms for Answers Validation task in Question Answering. Proceedings of Voronezh State University, Systems Analysis and Information Technologies 2, 181–188 (2011)
25. Toldova, S.J., et al.: NLP evaluation 2011–2012: Russian syntactic parsers. In: Computational Linguistics and Intellectual Technologies. Papers from the Annual International Conference "Dialogue", vol. 2(11), pp. 77–92 (2012)
26. Wang, R., Neumann, G.: Using Recognizing Textual Entailment as a Core Engine for Answer Validation. In: Advances in Multilingual and Multimodal Information Retrieval: 8th Workshop of the Cross-Language Evaluation Forum, CLEF 2007, Budapest, Hungary (2007)
27. Zhang, K., Shasha, D.: Simple fast algorithms for the editing distance between tree and related problems. SIAM J. Comput. 18(6), 1245–1262 (1989)

Author Index